COATED TEXTILES

COATED TEXTILES
Principles and Applications

A. K. Sen, M.Tech, Ph.D.

Emeritus Scientist
Defence Materials and Stores Research and
Development Establishment (DMSRDE)
Kanpur, India

TECHNICAL EDITOR

John Damewood, Ph.D.

Reeves Brothers, Inc.

TECHNOMIC
PUBLISHING CO., INC.
LANCASTER•BASEL

Coated Textiles
aTECHNOMIC®publication

Technomic Publishing Company, Inc.
851 New Holland Avenue, Box 3535
Lancaster, Pennsylvania 17604 U.S.A.

Printed in the United States of America
10 9 8 7 6 5 4 3 2 1

Main entry under title:
    Coated Textiles: Principles and Applications

A Technomic Publishing Company book
Bibliography: p.
Includes index p. 225

Library of Congress Catalog Card No. 2001088345
ISBN No. 1-58716-023-4

Table of Contents

Preface

COATED textiles applications are found in defense, transportation, healthcare, architecture, space, sports, environmental pollution control, and many other diverse end-product uses. I developed an insight into the breadth of the subject during my long association with the Defense Materials and Stores Research and Development Establishment (DMSRDE, Kanpur, India) while working on the development of protective clothing and related equipment. The opportunity to visit and work at several coating facilities has given me a feel for the complexity of the coated textile industry. The world production of coated fabrics used for defense alone every year is on the order of several billion dollars. Extensive research is being done on a global basis, and many new products, such as breathable fabrics, thermochromic fabrics, and charcoal fabrics, are entering the market. The subject is spread over a wide range of literature in polymer science and textile technology, with no single comprehensive book available. The motivation to write this book was to fill this void and to create a general awareness of the subject. This book is meant for scientists and technologists in academic institutions as well as in the coating and textile industry. The purpose of this book would be served if it could create additional interest in the coated textile industry and stimulate R&D activity to develop newer and better coated textile products.

A.K. Sen
Kanpur
2001

Acknowledgements

I had a long tenure with Defense Materials and Stores Research and Development Establishment (DMSRDE), Kanpur, India, as a research scientist, which has recently ended. DMSRDE, Kanpur, is a premier research establishment under the Defense Research and Development Organization (DRDO), Government of India, and is primarily responsible for research and development in nonmetallic materials required for the defense forces. The scope of R&D activity of the establishment encompasses a wide range of scientific disciplines including polymers, composites, organometallics, lubricants, anticorrosion processes, biodegradation studies, textiles and clothing, tentage, and light engineering equipment. It is one of the best equipped laboratories with various sophisticated analytical instruments.

The Department of Science and Technology (DST), Government of India, granted me a project to write a book on "Coated Textiles and their Applications." I am grateful to DST for this opportunity. I am also grateful to Professor G. N. Mathur, Director, DMSRDE, for permitting me to continue as Emeritus Scientist and for providing me access to a library and other facilities. His encouragement and help have been a source of inspiration. I am particularly thankful to my colleague Mr. N. Kasturia who first suggested that I write a book on the subject and for the help rendered at different stages in writing the book. I am indebted to Drs. V. S. Tripathi, L. D. Kandpal, Messrs. Anil Agrawal, T. D. Verma, Dhannu Lal, Darshan Lal, R. Indushekhar, G. L. Kureel, Miss Subhalakshmi, and several other colleagues at DMSRDE for their spontaneous help, suggestions, and input on specialized subjects. I also wish to acknowledge useful discussions and literature provided by Mr. A. K. Mody, Entremonde Polycoaters, Mumbai; Mr. M. L. Bahrani, Southern Group of Industries, Chennai; Mr. A. Narain, Swastik Rubbers, Pune; Mr. M. K. Bardhan, Director, SASMIRA, Mumbai; Professor P. Bajaj, I.I.T., Delhi; and Professor A. Nishkam, Principal, GCTI, Kanpur. Above all, I am thankful for the encouragement, inspiration, and valuable suggestions given by my wife, Sutapa.

Introduction

THE use of coated textiles for protective clothing, shelters, covers, liquid containers, etc., dates back to antiquity. Historically, the earliest recorded use of a coated textile was by the natives of Central and South America, who applied latex to a fabric to render it waterproof. Other materials such as tar, rosin, and wax emulsions have been used over the years to prepare water-resistant fabrics. Due to their vastly superior properties, rubber and other polymeric materials have become the preferred coatings. Today, coated fabrics are essentially polymer-coated textiles. Advances in polymer and textile technologies have led to phenomenal growth in the application of coated fabrics for many diverse end uses. Coated fabrics find an important place among technical textiles and are one of the most important technological processes in modern industry.

Textiles are made impermeable to fluids by two processes, coating and laminating. Coating is the process of applying a viscous liquid (fluid) or formulated compound on a textile substrate. Lamination consists of bonding a preprepared polymer film or membrane with one or more textile substrates using adhesives, heat, or pressure. Fibrous materials are also used for reinforcing polymeric materials to form composites for use in tires, conveyor belts, hoses, etc. The scope of this book has been restricted to coated and laminated textiles and does not address polymer fiber composites.

Several methods of production are used to manufacture a wide range of coated or laminated fabrics. Broadly, they are spread coating, dip coating, melt coating, and lamination. They not only differ in the processing equipment used, but also in the form of polymeric materials used. Thus, paste or solutions are required for spread coating; solutions are required for dip coating; and solid polymers such as powders, granules, and films are required for melt coating and lamination. The basic stages involved in these processes include feeding the textile material from rolls under tension to a coating or laminating zone, passing the coated fabric through an oven to volatilize the solvents and cure/gel the coating, cooling the fabric, and subsequently winding it up into rolls.

The properties of a coated fabric depend on the type of polymer used and its formulation, the nature of the textile substrate, and the coating method employed. The subject of coated textiles is thus interdisciplinary, requiring knowledge of polymer science, textile technology, and chemical engineering. The organization of this book is based on these considerations.

Among the various polymers used for coating and laminating, three classes are mainly used for coating: rubber, polyvinyl chloride, and polyurethane. These polymeric materials are specifically formulated with additives and compounded into a paste suitable for coating. The production of a polymeric coating fluid is one of the most important functions of the coating industry. The chemistry of these polymers, the additives used, and their processing for coating compounds and fluids have been described in Chapter 1. Conventional solvent coatings are losing favor, as they lead to environmental pollution. Several alternative processes, such as the ecofriendly aqueous polyurethane and radiation-cured coatings, are included in this chapter. The various adhesive treatments for improved elastomer-textile bonding have also been discussed in Chapter 1.

For many years, cotton was the primary fabric used for coating; however, today's coating industry uses diverse substrates made of rayon, nylon, polyester, polyester-cotton blends, and glass fibers that may be produced in woven, knitted, or nonwoven constructions. The physical properties of a coated fabric are affected by the nature of the fiber and the construction of the textile substrate. The choice of the substrate depends on the application of the material. Chapter 2 discusses the different fibers and their conversion into textile materials of various constructions used in the coating industry. The coating methods employed by the industry are discussed in Chapter 3. Emphasis is placed primarily on the principles, rather than on the engineering aspects, of the machinery. Chapter 4 describes the changes that occur in the physical properties of a fabric when it is coated. A brief account of rheological factors affecting the coating has been presented in Chapter 5. Thus, the raw materials, the coating methods, and the properties of the end product are presented in chronological sequence.

The large, ever-increasing variety of applications of coated fabrics is covered in the three subsequent chapters. Protective clothing for foul weather is one of the major applications of coated fabrics. In the last two decades, particularly after the development of GORE-TEX® laminates, there has been an explosion of development in breathable fabrics. Chapter 6 discusses all types of coated fabrics for foul weather protection with special emphasis on the developments in the field of breathable fabrics. Coated textiles used in synthetic leather, upholstery fabrics, fabrics for fluid containers, backcoating of carpets, and architectural textiles are discussed in Chapter 7.

In various applications of coated fabrics, a functional material such as dye, pigment, or carbon is applied on the textile materials with polymeric binders. These fabrics are being used as camouflage nets, thermochromic fabrics, protective clothing for toxic chemicals, etc. This specialized category of coated fabrics is included in Chapter 8. Metal coatings are finding newer uses in EMI-RFI shielding and radar responsive fabrics. These fabrics are also discussed in this chapter.

The test methods pertaining to coated fabrics have been discussed in Chapter 9. The references are given at the end of each chapter. Properties of common polymers used for coating are separately provided in Appendix 1, and a few typical formulations are given in Appendix 2.

CHAPTER 1

Polymeric Materials for Coating

1.1 RUBBER—NATURAL AND SYNTHETIC [1–4]

1.1.1 INTRODUCTION

IN ancient times, Mayan Indians waterproofed articles of clothing and footwear by applying (coating) gum from a tree (rubber tree) and drying it over smoke fires. Modern day history of coating rubber on fabrics dates to 1823, when the Scotsman Macintosh patented the first raincoat by sandwiching a layer of rubber between two layers of cloth [3].

Since then, there have been great advances in rubber-coated fabric technology. Coated fabrics are now used for diverse applications. Almost all types of rubbers are used for coating, but the discussion here will be restricted to the more popular kinds.

Rubber is a macromolecular material that is amorphous at room temperature and has a glass transition temperature, T_g, considerably below ambient. Raw rubber deforms in a plastic-like manner, because it does not have a rigid network structure. It can be cross-linked by vulcanization to form an elastomer with the unique ability to undergo large elastic deformations, that is, to stretch and return to its original shape. For natural and most synthetic rubbers, vulcanization is accomplished with sulfur.

Elastomers that have stereoregular configuration and do not have bulky side groups or branching undergo crystallization. Crystallization cannot occur above melt transition, T_m. The rate of crystallization is greatest at about halfway between T_g and T_m. For natural rubber, for example, this is about $-25°C$. The crystallites embedded in the elastomeric matrix act as physical cross-links, like reinforcing fillers. Most importantly, crystallinity can be induced by stress. Formation of crystallites enhances the strength of the rubber.

Vulcanization lowers the crystallinity as three-dimensional networks create obstacles in segments entering the crystal lattice. Lower crystallinity is also observed in random copolymers.

1

1.1.2 PRODUCTION, STRUCTURE, AND PROPERTIES

1.1.2.1 Natural Rubber (NR)

NR is obtained from the exudation of the rubber plant, *Hevea brasiliensis*. The rubber is obtained from the latex by coagulation, sheeting, drying, and baling. There are various internationally recognized market grades, common among them are ribbed smoked sheets and pale crepe. Natural rubber contains about 90% rubber hydrocarbon as *cis*-1,4-polyisoprene along with naturally occurring resins, proteins, sugars, etc., that precipitate during latex coagulation. The average molecular weight of polyisoprene in natural rubber ranges from 200,000 to 500,000, with a relatively broad molecular weight distribution. As a result of its broad molecular weight distribution, NR has excellent processing behavior.

$$-CH_2\underset{\displaystyle CH_3}{\overset{\displaystyle }{C}}=C\underset{\displaystyle CH_2}{\overset{\displaystyle H}{}}-CH_2\underset{\displaystyle CH_3}{\overset{\displaystyle }{C}}=C\underset{\displaystyle CH_2}{\overset{\displaystyle H}{}}-$$

cis-1,4-Polyisoprene unit

The α-methylene group of the polyisoprene units is reactive for vulcanization with sulfur. NR vulcanizates combine a range of properties that are of great technological interest. The individual property can be improved by the use of synthetic rubber, but a combination of high tensile strength, resilience, dynamic properties, and good low temperature flexibility make NR indispensable for several applications. The high tensile strength and tear resistance of NR vulcanizates is due to strain crystallization. Being nonpolar, NR swells in nonpolar solvents. Reaction of the double bond in the polyisoprene unit with oxygen or ozone results in degradation of the polymer.

1.1.2.2 Styrene-Butadiene Rubber (SBR)

SBR is a copolymer of styrene and butadiene. The styrene content ranges from about 25 to 30 wt.%. The structure is given as

$$-(CH_2-CH=CH-CH_2)_x-(CH_2-\underset{\displaystyle \bigcirc}{CH})_y-$$

SBR

SBR is mainly prepared by emulsion polymerization, The monomers are randomly arranged in the chain, and the butadiene part is mainly in the *trans* configuration ($\sim$75%). Some 1,2-addition products are also formed. Depending on the temperature of polymerization, SBR may be classified into hot-polymerized and cold-polymerized grades. The hot grades are highly branched compared to the cold grades.

SBR can also be copolymerized in a solution process using alkyl lithium catalysts in a nonpolar solvent. These rubbers have much higher *cis*-1,4-butadiene content (50–55%), less chain branching, and narrower molecular weight distribution.

$$-CH_2 \quad CH_2- \qquad -CH_2 \qquad\qquad -CH_2-CH-$$
$$C=C \qquad\qquad CH=CH \qquad\qquad\qquad CH$$
$$\qquad\qquad\qquad\qquad CH_2- \qquad\qquad\qquad CH_2$$

cis	*trans*	
1,4-Butadiene		1,2-Addition product

SBR does not crystallize even on stretching their vulcanizates, therefore, pure gum strength is generally low. It has better heat and aging resistance than NR and is usually used in combination with NR and other rubbers.

1.1.2.3 Isoprene-Isobutylene Rubber, Butyl Rubber (IIR)

Butyl rubber is a copolymer of 97 to 99.5 mole % of isobutylene and 0.5 to 3 mole % isoprene. The isoprene unit provides the double bond required for sulfur vulcanization.

$$
\begin{array}{ccc}
CH_3 & & CH_3 \\
| & & | \\
-(CH_2 - C -)_x(- CH_2- C = CH\text{-}CH_2)_y- \\
| \\
CH_3
\end{array}
$$

IIR, Butyl rubber

It is produced by cationic polymerization in methylene chloride with $AlCl_3$ as catalyst, at subzero temperatures ($-90°C$ to $-100°C$). The isobutylene monomer units polymerize mainly in head-to-tail arrangements, and the isoprene units in the polymer chains polymerize in *trans* 1,4-configuration. The molecular weight ranges between 300,000 to 500,000. On halogenation of IIR in an inert organic solvent, a rapid electrophilic substitution takes place, and one halogen atom is substituted per isoprene unit, mainly in the allylic position. Thus, a small number of halogen atoms are incorporated into the polymer chain. These are known as chlorobutyl rubber (CIIR) or bromobutyl rubber (BIIR) depending on the halogen substituted. The polymer chains are highly saturated and have a very regular structure due to the symmetrical nature of the monomer. As a result, butyl rubber exhibits very low gas permeability, ozone, heat, weathering, and chemical resistance. Butyl rubbers are self-reinforcing with a high gum tensile strength. The halobutyl rubbers cure faster than butyl rubber. Bromobutyl has much lower gas permeability and better resistance to aging, weathering, and heat than butyl rubber. Butyl and bromobutyl rubbers are especially used where low gas permeability is required.

1.1.2.4 Ethylene Propylene Polymers (EPM) and Terpolymers (EPDM)

Copolymers of ethylene and propylene EPM are made by solution polymerization using vanadium containing alkyl aluminum, Ziegler-type catalyst. These are elastomers, but they do not contain any double bonds.

$$-(CH_2\text{-}CH_2)_x\text{-}(\underset{\underset{CH_3}{|}}{C}H\text{-}CH_2)_y\text{-}$$

EPM

For good elastomeric properties, the ethylene propylene ratio ranges from 45–60 wt.%, and the monomers are arranged randomly. Consequently, these polymers are predominantly amorphous, and the pure gum strength is low. The molecular weight ranges between 200,000 to 300,000. Because EPM is saturated, the polymeric chain is cured by peroxides. The terpolymer EPDM contains, in addition to the olefin monomers, a nonconjugated diene as the third monomer, which renders EPDM able to be vulcanized by sulfur. The common third monomers are dicyclopentadiene, ethylidene norbornene, and 1,4-hexadiene. In these dienes, one double bond is capable of polymerizing with the olefins, but the other is not a part of the main chain.

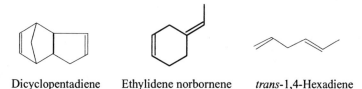

Dicyclopentadiene   Ethylidene norbornene   *trans*-1,4-Hexadiene

Both EPM and EPDM have excellent resistance to oxygen, ozone, heat, and UV radiation.

1.1.2.5 Polychloroprene Rubber (CR)

These rubbers are produced by the emulsion polymerization process of 2-chloro-1,3-butadiene. The polymer chains consists of approximately 98%

1,4-addition products which are mostly *trans* in configuration, and the rest are 1,2-addition products. The 1,2-addition product contains a chlorine atom attached to a tertiary allylic carbon atom that is highly activated and thus becomes the curing site in the polymer chain.

| *trans*-Polychloroprene unit | 1,2-Addition product |

The temperature of polymerization has an important bearing on the polymer structure. At higher temperatures, there is less uniformity in the chain due to large proportions of 1,2 and 3,4 moieties and other isomers in the monomeric sequences. On the other hand, at lower temperatures, the polymeric chain is more regular. The CR is also available in sulfur-modified grades. The polymer chains have Sx groups, and this aids processing due to easy depolymerization. Unlike the diene rubbers, CRs are not vulcanized by sulfur but are vulcanized by metal oxides—a combination of MgO and ZnO.

Polychloroprene stiffens at low temperatures. This is due to second-order transition and crystallization. The rate of crystallization is most rapid at $-10°C$. Though the stiffening is reversible, it is detrimental for the production of certain goods. Incorporation of low temperature plasticizers like butyloleate can lower the stiffening temperature of CR compounds. CR produced by high temperature polymerization has a much lower rate of crystallization than that produced at a low temperature. The high crystallizable grades are useful as adhesives. For production of coated fabrics, materials with long crystallization times are chosen, as softness and flexibility are more important than ability to withstand heavy stress. The gum vulcanizates of CR show high tensile strength because of strain crystallization, but the resilience is lower. Polychloroprene rubbers are resistant to oxidation, ozone degradation, and flex cracking, and, because of the chlorine atom in the molecule, are inherently flame resistant. Because of its polar nature, the rubber is resistant to hydrocarbons, fats, oils, and most chemicals. It is used for applications requiring weather, oil, ozone, and flame resistance.

1.1.2.6 Nitrile Rubber (NBR)

NBR is a copolymer of acrylonitrile and butadiene obtained by emulsion polymerization. The acrylonitrile content varies from 18–50%, depending on the properties desired. The molecular weight ranges from 20,000 to 100,000.

$$-(-CH_2 - CH = CH - CH_2)_x - (CH_2 - CH)_y--$$
$$|$$
$$CN$$

NBR

Like SBR, varying temperatures of polymerization produce different grades of NBR. NBR produced at low temperatures shows less branching than hot rubbers. The steric configuration, i.e., *cis*-1,4, *trans*-1,4, and *trans*-1,2 structures are also influenced by polymerization temperature. The lack of compositional uniformity along the polymer chains prevents formation of crystallites on extension. This results in poor tensile properties of NBR gum vulcanizates.

Nitrile rubbers are of special interest because of their high degree of resistance to fuels, oils, and fats. An increase in acrylonitrile (AN) content increases its oil resistance because of enhancement of polarity of the rubber. NBR has a low gas permeability. Increase of AN percentage in NBR lowers its gas permeability but adversely affects its low temperature flexibility and resilience. NBR is extensively used where oil resistance is required.

1.1.2.7 Chlorosulfonated Polyethylene Rubber (CSM)

CSM is produced by reaction of polyethylene solution with chlorine and sulfur dioxide in the presence of UV radiation.

$$--CH_2 - CH - CH_2 - CH_2 - CH_2 - CH---$$
$$|$$
$$Cl$$
$$SO_2Cl$$

CSM unit

Commercial grades contain 25–40 wt.% of chlorine and about 1% of S. The chlorine and sulfur are randomly distributed along the polymer chain. It is cross-linked by metal oxides through chlorine atom and chlorosulfonyl group. These rubbers are characterized by a unique combination of special properties like ozone resistance, flame retardance and resistance to corrosive chemicals and oxidizing agents.

1.1.2.8 Silicone Rubber

Silicone rubber is obtained from silicones. Chemically, silicones are polysiloxanes containing Si-O- bonds. The most important polymers are polydimethyl siloxane, polymethyl phenyl siloxane, and vinyl methyl siloxane. They are manufactured by the hydrolysis of the appropriate dichlorosilane R_2SiCl_2. Silicones are available in a wide range of molecular weights and viscosities, from fluids to gums.

$$\begin{array}{ccc}
CH_3 & CH_3 & CH_3 \\
| & | & | \\
-O-Si--O- & -O--Si--O- & --O--Si--O-- \\
| & | & | \\
CH_3 & C_6H_5 & CH \\
& & \parallel \\
& & CH_2
\end{array}$$

Dimethyl siloxane unit   Methyl phenyl siloxane unit   Vinyl methyl siloxane unit

There are three major ways of curing silicone rubbers, viz., peroxide cure, hydrosilation, and condensation cure [5,6].

(1) Peroxide initiated cure: $ROOR \xrightarrow{\Delta} 2R\dot{O}$

$$\begin{array}{ccccc}
CH_3 & CH_3 & RO. & CH_3 & CH_3 \\
| & | & & | & | \\
--O-Si-CH =CH_2 & + CH_3-Si-O- & \rightarrow & --O-Si--CH_2-CH_2-\dot{C}H-Si-O- & --\rightarrow \text{cross-link} \\
| & | & & | & | \\
O & O & & O & O \\
| & | & & | & |
\end{array}$$

$$(1)$$

This is a one-part system where the peroxide is activated on heating above 100°C. The vinyl group facilitates free radical reaction with the formation of vinyl to methyl and methyl to methyl bonds.

(2) Hydrosilation: an addition reaction occurs between vinyl siloxane and siloxane containing Si-H group catalyzed by Pt (chloroplatinic acid). The cross-link occurs due to multiple functionality of both reactants. This is a two-part reaction with one part containing vinyl siloxane with Pt catalyst and the other containing silicone with Si-H functionality. The two liquid parts permit direct processing including coating without solvent and are known as liquid silicone rubber LSR.

$$\begin{array}{ccccc}
CH_3 & CH_3 & Pt & CH_3 & CH_3 \\
| & | & & | & | \\
-O-Si-CH=CH_2 & + H-Si -O- & -----\rightarrow & -O-Si-CH_2--CH_2-Si-O- & --\rightarrow \text{cross link} \\
| & | & & | & | \\
O & O & & O & O \\
| & | & & | & |
\end{array}$$

PDMS with Si-H group

$$(2)$$

(3) Condensation cure: condensation reaction between siloxanes with terminal hydroxyl groups, -silanols, and a cross-linker,–tri or tetra functional organo silicon compound, leads to cross-linking/cure.

$$
\begin{array}{ccccccc}
R & R & R & X & -HX & R & X & R \\
| & | & | & | & & | & | & | \\
\text{HO-Si-(O -Si-O)}_n\text{-Si-OH} & + & \text{X-Si-X} & \rightarrow & \text{-Si-(O-Si-O-)}_n\,\text{Si-O} & \dashrightarrow \text{cross-link} \\
| & | & | & | & & | & | & | \\
R & R & R & X & & R & X & R \\
\end{array}
$$

    Silanol   Cross-linker

(3)

Common cross-linking agents are tetraethyl ortho silicate ($X = OEt$), triacetoxy silane ($X = OAc$), etc. Condensation occurs at room temperature in the presence of metal soap catalysts, typically, stannous octoate and moisture. The rubbers are known as room temperature vulcanizates (RTV). The RTVs may be two-part systems or one-part systems. In a one-part system, the common cross-linker is methyl triacetoxy silane, and the reactants are stored in anhydrous condition.

The molecular weights of hot vulcanizates range from 300,000 to 1,000,000. The Si-O bond energy is higher than that of the C-C bond, as such, the polysiloxane chain is thermally and oxidatively much more stable than organic hydrocarbon chains. Silicone rubbers can also retain their flexibility to as low as $-100°C$. The stability of silicones over a wide range of temperatures is outstanding and is not found in any other rubber. Silicones are extraordinarily resistant to aging, weathering, and ozone. They have, however, lower mechanical properties, but they do not change much with temperature. The vulcanizates are hydrophobic and are resistant to chemicals. They form transparent/translucent coatings. Because of their unique properties, they find specialized applications, for example, gaskets, O-rings, wire cable, etc.

1.1.3 RUBBER ADDITIVES

Various chemicals, fillers, accelerators, and cross-linking agents are added to rubber to facilitate processing and vulcanization. The properties of the end product can be significantly altered by proper choice and amount of the compounding ingredients to meet the diverse end-use requirements.

Typical ingredients of a rubber formulation are as follows:

- raw rubber
- cross-linking agents
- accelerators
- accelerator activators

- antidegradants
- fillers, reinforcing, diluents
- processing aids
- pigments and dyes
- special additives, e.g., flame retardants, fungicides

The cross-linking agent for natural rubber is sulfur along with organic accelerators. Zinc oxide and stearic acid are used as activators of the accelerator. The synthetic olefin rubbers like SBR, NBR, butyl, and EPDM can also be cured by sulfur system. For butyl and EPDM rubbers having low unsaturation, more active accelerators are used with a higher temperature of cure. Apart from elemental sulfur, organic compunds that liberate sulfur at the temperature of vulcanization can also be used for vulcanization, e.g., dithiodimorpholine and 2-morpholine-dithio benzothiazole.

The nonolefin rubbers, polychloroprene and chlorosulfonated polyethylenes are cured by metal oxides. In the case of polychloroprene, the curing is done by zinc oxide and magnesium oxide, because sulfur cure is not possible as the double bond is hindered by the neighboring chlorine atom. The zinc oxide reacts with chlorine atoms of the 1,2 units present in the chain after an allylic shift. Magnesia also acts as scavenger of the chlorine atom. More rapid cure is achieved by the use of organic accelerators like ethylene thiourea. Chlorosulfonated polyethylene is cured by litharge and magnesia, along with accelerators. The curing occurs by ionic and covalent bond formation through sulfonyl chloride groups.

In the case of ethylene propylene rubber (EPM) and silicones, peroxides are the cross-linking agents. In EPM, the tertiary carbon atom at the point of branching is attacked by peroxides with H abstraction, generation of a tertiary free radical, followed by termination, forming a cross-link. The cross-linking of polydimethyl siloxane by peroxides, as discussed earlier, occurs through abstraction of H from the methyl group with the formation of free radical followed by termination. More rapid vulcanization is achieved when vinyl groups are present in the chain. Some common peroxides are benzoyl peroxide, 2,4-dichlorobenzoyl peroxide, dicumyl peroxide, etc.

The accelerators used for sulfur vulcanization of olefin rubbers are of various types depending on the rate of cure, i.e., medium, semi-ultra, and ultra accelerators. The chemical types are aldehyde amine, guanidines, thiazoles, sulfenamides, dithiocarbamates, thiuram sulfides, and xanthates. The action of the accelerator is further enhanced by activators. Zinc oxide and stearic acid system for sulfur vulcanized rubbers is most common, but other zinc salts of fatty acids like zinc laurate can be used. Most of the rubber formulations contain antidegradants, antioxidants, and antiozonants. They function by either capturing the free radical formed during the degradation process or by decomposing the peroxides and hydroperoxides produced into nonreactive fragments. The majority of the commercially available inhibitors belong to two main chemical classes: amines and phenolics. Fillers are incorporated into a rubber formulation for reinforcement, i.e., for enhancement of tensile strength, abrasion resistance, and tear resistance, or as diluent to reduce cost. The most common reinforcing filler is carbon black. Among the non-black reinforcing fillers, common are precipitated silica, fume silica, calcium silicate, hydrated aluminium silicate (clay), etc. Barytes, whitings, talc, chalk, kaolin, and kieselguhr are some of the other fillers used. Plasticizers and softeners are added to rubber compounds to aid various processing operations of mixing, calendering, extruding, and molding. These include a wide range of chemicals such as petroleum products, oils, jelly, wax, coal tar products, pine tar, fatty acid salts, factice (reaction product of vegetable oil and sulfur), and esters of organic acids. Peptizers are also added to increase the efficiency of molecular breakdown, facilitating the mastication process. Pentachlorophenol, its zinc salt, and di-(o-benzamido phenyl) disulfide, are common peptizing agents. For noncarbon black rubber compounds, coloring materials are used, which are generally colored inorganic compounds. Sometimes, special purpose additives are added to obtain specific properties, like blowing agents for cellular rubber, flame retardants, like chlorinated paraffins, zinc borate, etc.

1.1.4 COMPOUNDING AND PROCESSING OF RUBBER

For coating of rubber on fabric, it has to be properly processed. The steps involved are as follows:

- mastication or milling
- compounding
- coating by calendering
- preparing dough and spread coating
- vulcanizing the coated fabric

The processing and machinery required for rubber are different from those required for coating other polymers. A brief account of the processing steps are discussed here.

1.1.4.1 Mastication and Compounding

Raw rubber is masticated to decrease its viscosity to a desired level for incorporation of the compounding ingredients and their proper dispersion. A proper adjustment of viscosity is also required for various processing operations. When rubber has all of the ingredients needed, it is known as a "compound." If some ingredients have been deliberately withheld, particularly curing agents, the partially completed compound is known as the master batch. Mastication is done generally in mixing mills or in internal mixers.

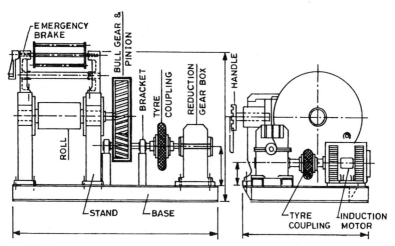

Figure 1.1 Line diagram of a two-roll mixing mill. Courtesy M/s Slach Hydratecs Equipment, New Delhi, India.

Mixing mills are used for small-size compounds and as a follow-up to internal mixers. They consist of two horizontal and parallel rolls made from hard castings, that are supported through strong bearings in the mill frame. The distance between the rolls is adjustable. The two rolls move in opposite directions at different speeds, the back roll running faster than the front (Figure 1.1). The ratio of the two speeds is known as the friction ratio, which is about 1:1.25 for natural rubber. Frictioning promotes tearing, kneading, and mixing of the rubber mass and the ingredients in the roll nip. Side guides are provided to prevent rubber from flowing to the bearings. The rolls can be cooled or heated with water or steam by circulating water through a drilled core or through peripheral holes. Raw rubber is first placed between the rolls, the elastomer is torn, and then it wraps around the front roll. After several passages, a continuous band is formed. The degree of mastication is controlled by the temperature of the roll, the size of the nip, and the number of passes. The compounds are then added in a well-defined sequence, such as accelerators, antioxidants, factice, pigments, fillers, and sulfur. If peptizers are used, they are added first. At the end, the compounded layer is cut off repeatedly in order to homogenize it.

The mixing in an internal mixer is done in a closed chamber by rotating kneading rolls. The mixer can handle larger batches. The internal mixer consists of a mixing chamber shaped in the form of a horizontal figure of eight. Two rotors are fitted into the chamber that rotate at different speeds to maintain a high friction ratio. The walls of the mixing chamber and the rotors are equipped with cavities for cooling or heating. The mixing chamber has a filling device at the top, which is used for adding ingredients.

The chamber is closed by a pneumatically operated ram to ensure that the rubber and ingredients are in proper contact. A ram pressure of about 2–12 kg/cm^2

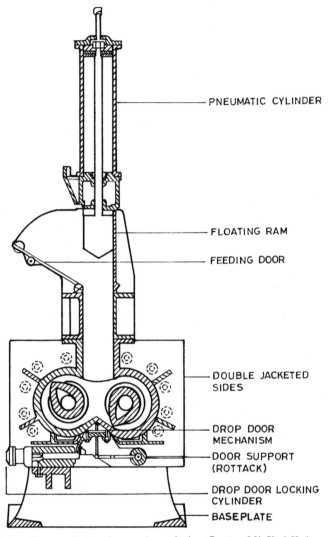

PNEUMATIC CYLINDER

FLOATING RAM

FEEDING DOOR

DOUBLE JACKETED SIDES

DROP DOOR MECHANISM

DOOR SUPPORT (ROTTACK)

DROP DOOR LOCKING CYLINDER

BASEPLATE

Figure 1.2 Line diagram of a drop door type internal mixer. Courtesy M/s Slach Hydratecs Equipment, New Delhi, India.

is applied for this purpose. Higher ram thrust results in better mixing efficiency. The design of the rotors may be (1) tangential type, where the kneading action is between the rotors and the jacket or (2) intermeshing type, where the kneading occurs between the rotors. A drop door type internal mixer is shown in Figure 1.2.

The rotor speed varies between 20–60 rpm. The capacity of the internal mixer can vary widely, but mixers of 200 kg batch size are quite popular. The operation of the mixer consists of adding rubber in split or pelletized forms. After a short mastication, the compounding ingredients are added. The incorporation of rubber and other ingredients can be partly or fully automated. Normally, complete mixing is not done in an internal mixer because of the scorch problem. Master batch is initially prepared and the remaining ingredients are added to it afterwards in a mixing mill. The compound coming out of the internal mixer is in the form of lumps and has to be cooled and homogenized in sheeting mills. Mixing can be done continuously in a single- or twin-screw extruder, however, extruders are not very popular for mixing rubbers.

1.1.4.2 Spreading Dough

Preparation of dough of the right consistency is of great importance for spread coating. The compounded rubber is cut thinly in narrow strips or pieces and soaked in a proper solvent for a few hours. Organic solvents capable of dissolving the elastomers are selected for the purpose, keeping the cost factor in view. Mixed solvents are also used quite often for better solvating properties, enabling preparation of a more concentrated solution. Toluene, aromatic and chlorinated hydrocarbons, and esters are common solvents used. For natural rubber, toluene is usually used. The soaked mass is then transferred into solution kneaders for preparation of a homogenous dough. The solution kneaders consist of a semicircular trough with a lid and two kneading paddles in the shape of Z or sigma. The paddles turn with a differential speed, the forward one is 1.5 to 2 times faster than the back. The trough is double walled to permit cooling and is closed by a movable lid. The rotating paddles disintegrate the swollen mass while continuously wetting the rubber. The agitation is continued until a dough of the right consistency is produced; this may take up to 12 hrs. The dough is emptied from the trough either by tilting or by discharge screws as per the design of the kneader. However, high speed Ross mixers do this job in 30 min to 1 h.

1.1.4.3 Vulcanization of Rubberized Fabrics

Vulcanization can be carried out as a batch process in a steam autoclave using saturated steam. An autoclave is a cylindrical pressure vessel, normally used in the horizontal position. Curing can also be done by hot air under pressure. However, because the heat transfer coefficient of air is lower than steam, air curing requires higher curing time, needing a change in formulation. Moreover, the oxygen of air can oxidize the elastomer. In steam curing, on the other hand, formation of condensed water can lead to unsightly water spots, and local undercuring. This problem is solved by coating the uncured article with a wetting agent.

Continuous drum cure, also known as rotocure, is also used. In this process, the curing of the coated fabric is achieved by placing the fabric in contact with a rotating steam-heated drum (Figure 1.3).

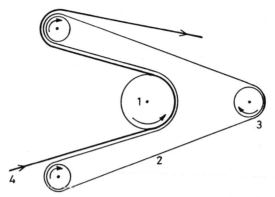

Figure 1.3 Rotocure process: (1) steel drum, (2) steel belt, (3) tension roll, and (4) fabric.

A steel band runs over about two-thirds of the circumference of a slowly rotating steam-heated steel drum. The steel band is pressed against the drum. The sheet to be vulcanized runs between the drum and the steel band and is pressed firmly against the drum with a pressure of about 5–6 kg/cm^2. The sheet slowly moves with the rotation of the drum. Vulcanization occurs because of the temperature of the drum and the pressure on the sheet created by the steel band.

1.2 POLYVINYL CHLORIDE [7–11]

1.2.1 INTRODUCTION

Polyvinyl chloride (PVC) is one of the few synthetic polymers that has found wide industrial application. The popularity of PVC is due its low cost, excellent physical properties, unique ability to be compounded with additives, and usefulness for a wide range of applications and processability by a wide variety of techniques. The repeat unit of PVC is

$$--CH_2-CH--$$
$$|$$
$$Cl$$

The units are linked mainly head to tail, with very few head to head links. PVC is considered to be an amorphous polymer. The crystallinity is only about 10%. This is attributed to the nonregular position of the chlorine atoms around the carbon chain. Branching is low in PVC. Lower polymerization temperature favors more linear structure. The molecular weight ($\bar{M}_n$) of commercial resins, ranges from 50,000 to 100,000.

It is produced by the addition polymerization of vinyl chloride, $CH_2=CH\text{-}Cl$. The methods of polymerization are suspension, emulsion, mass, and solution processes. Among these, suspension is the most favored commercial process, contributing to about 80% of total polymer production. The emulsion and mass processes contribute to about 10% each. The resins obtained from suspension and mass polymerization are porous and absorb plasticizer rapidly. Polymer obtained by the emulsion process is nonporous, with very fine particle size. It quickly and reversibly absorbs plasticizer once heated at temperatures above 80°C. The main use of emulsion PVC is in plastisol and organosol preparations, which are extensively used in coating and slush molding. The solution process is almost exclusively restricted to the manufacture of PVC copolymers for use in surface coatings.

1.2.2 RESIN CHARACTERISTICS

The physical forms of PVC resins are diverse which permits a wide range of processing techniques. The resins are classified on the basis of important properties required for processing. The specifications of the resins have been standardized by ASTM and ISO; and a nomenclature system to designate the properties evolved. The characteristics of general purpose resins (G) and dispersion resins (D) are different.

The important characteristics of general purpose resins are as follows:

(1) Molecular weight: in industrial practice, dilute solution viscosity is normally determined as an index of molecular weight. The results are commonly expressed in terms of K value or viscosity number. The $\bar{M}_v$ can be determined from the Mark-Houwink equation. A high K value denotes high molecular weight, a high melt viscosity of the unplasticized PVC requiring higher processing temperature.

(2) Particle size: particle size and particle size distribution influence compounding and processing properties.

(3) Bulk density

(4) Dry flow: this property is a measure of the ease of handling of granular resins.

(5) Plasticizer absorption: it is a measure of the capacity of the resin to absorb plasticizer, yet remain a free-flowing powder. It is dependent on surface properties of the resin powder.

(6) Electrical conductivity: this test is intended to distinguish between electrical and nonelectrical grades.

Important characteristics for dispersion resins are as follows:

(1) Molecular weight

(2) Particle size

(3) Settling

(4) Plastisol viscosity: usually determined by viscometer at specified shear rates with a specified concentration of plasticizer (dioctyl phathalate)

(5) Plastisol fusion: it denotes the complete solvation of the resin by the plasticizer and is a function of solvating power of plasticizer, temperature, and time. Determination of clear point and measurement of rate of increase of viscosity at a constant temperature are some of the methods of determination of fusion.

1.2.3 ADDITIVES FOR PVC

For processing and imparting properties for special applications, PVC is compounded with a variety of additives. Some of the important additives are as follows:

- plasticizers
- heat stabilizers
- fillers
- lubricants
- colorants
- flame retardants

1.2.3.1 Plasticizers

Plasticizers are an important additive of PVC resin, because the majority of PVC products are plasticized. These are liquids of low or negligible volatility or low molecular weight solids, which when incorporated into the polymer, improve its processability and impart end product softness, flexibility, and extensibility. The other concomitant effects of plasticization are lowering of T_g and softening temperature, reduction of strength, and increased impact resistance. The plasticizer acts by lowering the intermolecular forces between the polymer chains. The plasticizer should be compatible with the polymer, or exudation will occur. Those plasticizers that are highly compatible with PVC are known as primary plasticizers, while plasticizers that have limited compatibility are known as secondary plasticizers. Secondary plasticizers are added to impart special properties or to reduce cost.

1.2.3.1.1 Plasticizer Characteristics

The main parameters for ascertaining the effectiveness of a plasticizer are compatibility, efficiency, and permanence.

(1) Compatibility: this can be determined from solubility parameter δ and Flory Huggins interaction parameter χ. The PVC-plasticizer system is considered

compatible if its solubility parameters are nearly equal. Again, if the χ value is low (<0.3), the system is considered compatible. Compatibility can also be determined from clear point, which is the temperature at which the PVC-plasticizer mixture becomes clear. The lower the clear point temperature, the greater the compatibility.

(2) Efficiency: technologically, it is the amount of plasticizer required to produce a selected property of practical interest, like hardness, flexibility, or modulus. The efficiency of plasticizer can also be gauged by the lowering of T_g, and changes in dynamic mechanical properties.

(3) Permanence: the plasticizer may be lost from the compounded resin by vaporization into the atmosphere, extraction in contact with a liquid, or migration into a solid in intimate contact with the plasticized PVC. Permanence can be determined by weight loss measurements on exposure to the extraction media.

1.2.3.1.2 Plasticizer Types

(1) Phthalates: these are the largest and most widely used group of plasticizers. The esterifying alchohol ranges from methyl to tridecyl. The lower chain length esters have high solvating power but suffer from high volatility and poor low temperature properties. Medium chain C_8 phthalates possess optimum properties. The longer chain C_{10}–C_{13} esters have reduced solvating power and efficiency, though low volatility. Di-2-ethylhexyl phthalate (DOP) and diisoctyl phthalate (DIOP) are extensively used in industry because of their better balance of properties.

(2) Phosphates: these are organic esters of phosphoric acids. The triaryl phosphates, like tricresyl (TCP) and trixylyl (TXP), are by far the most important phosphate plasticizers.The triaryl phosphates offer excellent flame retardance, good solvating power, and good compatibility, but poorer low temperature properties.

(3) Aliphatic diesters: in this category are esters of adipic, azelaic, and sebacic acids of branched chain alcohols such as isooctanol, 2-ethylhexanol, or isodecanol. These impart low temperature flexibility to PVC compositions. Their compatibility is, however, low, and they are categorized as secondary plasticizers.

(4) Epoxies: epoxidized soybean oil and linseed oil exhibit good plasticizing and stabilizing actions. They possess low volatility and good resistance to extraction.

(5) Polymeric plasticizers: the majority of commercial plasticizers of this class are saturated polyesters, synthesized by the reaction of a diol and dicarboxylic acid along with an end capping agent, which may be a monohydric alcohol or monocarboxylic acid. An increase in molecular weight results

in improved permanence and lower volatility, but it adversely affects low temperature properties and compatibilty.

1.2.3.2 Heat Stabilizers

Unless suitably protected, PVC undergoes degradation at the processing temperatures. The manifestations of the degradation include the evolution of hydrogen chloride, development of color from light yellow to reddish brown, and deterioration of mechanical properties. The degradation occurs due to the progressive dehydrochlorination of the polymer chains with the formation of conjugated double-bond polyenes, possibly by a free radical mechanism. The site of initiation could be a chlorine atom attached to a tertiary carbon atom at the site of branching. As shown in Figure 1.4, the first step is the formation of an allylic group, whose Cl atom is strongly activated by the neighboring double bond favoring further elimination of HCl. The HCl acts as an autocatalyst.

In addition to the polyene formation, the polymer undergoes chain scission, oxidation, cross-linking, and some cyclization. The Diels Alder reaction between polyene moieties of neighboring chains is believed to be responsible for much of the cross-linking. PVC also undergoes photodegradation on exposure to light in the presence as well as in the absence of oxygen. The manifestations and degradation products are similar to those of thermal degradation. In plasticized PVC, exudation of plasticizers from PVC occurs on weathering, which has been attributed to their partial exclusion from the areas where cross-linking has occurred. A heat stabilizer should prevent the reaction responsible for degradation of PVC. It should bind the liberated hydrogen chloride, deactivate potential initiation sites by substituting stable groups for labile chlorine, disrupt formation of polyene sequence, and deactivate the free radicals. Morever, the stabilizer should be compatible with the polymer and other additives in the compound.

The heat stabilizers may be classified as lead compounds, organo tin compounds, compounds of other metals like barium, cadmium, and zinc, and organic

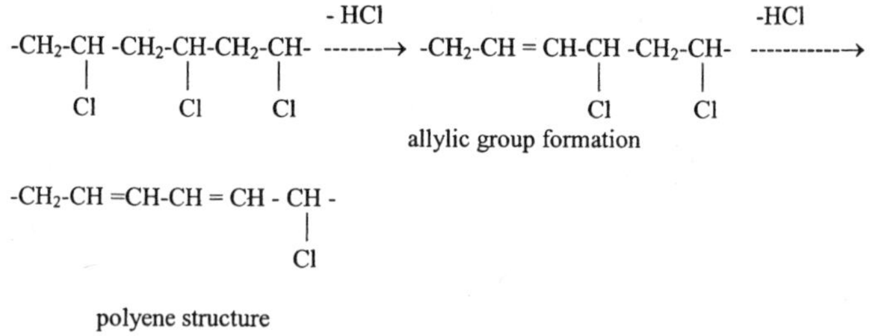

$$-CH_2-CH-CH_2-CH-CH_2-CH- \xrightarrow{-HCl} -CH_2-CH=CH-CH-CH_2-CH- \xrightarrow{-HCl}$$

with Cl, Cl, Cl substituents and allylic group formation (Cl, Cl)

allylic group formation

$$-CH_2-CH=CH-CH=CH-CH-$$

with Cl substituent

polyene structure

Figure 1.4 Formation of polyene structure.

compounds. Common lead and tin compounds are lead phosphate, dibasic lead stearate, dibutyl tin laurate, dibutyl tin maleate, etc. Barium, cadmium, and zinc salts, when used in combination, impart excellent stability and show synergistic effect. The compounds are salts of fatty acids, like laurates, stearates, and octoates. These compounds are used in calendered goods, plastisols, floorings, and coated fabrics. Organic stabilizers are epoxidized oils, phosphites, and polyhydric alcohols. They are normally regarded as secondary stabilizers in conjunction with Ba-Cd-Zn stabilizers.

For some special applications antioxidants are added to the PVC compound. They are usually phenolics like 2,6-di-t-butyl-4-methyl phenol and 3-(3,5-di-t-butyl-4-hydroxy phenyl) octadecyl propionate. The degradation of PVC by UV radiation can be prevented by incorporation of UV absorbers like derivatives of 2-hydroxy bezophenone, benzo triazoles, etc., or by addition of inorganic particulate screening agents like carbon black and titanium dioxide.

1.2.3.3 Other Additives

(1) Fillers: the primary role of a filler in PVC is to reducte cost, but they can play a functional role by improving processing and properties of the end product. The common fillers are (1) calcium carbonate fillers—whiting, and marble dust, (2) silicates—clay, talc, and asbestos, and (3) barytes.

(2) Lubricants: the role of a lubricant is to facilitate processing and control the processing rate. Mineral oil, silicone oils, vegetable oils, and waxes are common lubricants. Metal stearates of Pb, Ba, Cd, and Ca may be used for the dual purpose of stabilizing and lubricating. The compatibility of lubricants is low, resulting in their exudation at processing conditions.

(3) Colorants: the colorants of PVC are inorganic and organic pigments. The inorganic pigments include titanium dioxide, chromium oxide, ultramarine blue, molybdate orange, etc. The organic pigments are phthalocyanines, quinacridines, and benzidines. The inorganic pigments have excellent heat resistance, light stability, and opacity.

(4) Flame retardants: the inherent flame retardant property of PVC due to the presence of a chlorine atom is affected by the addition of flammable plasticizers. Antimony trioxide and borates of zinc and barium are widely used for this purpose. Chlorinated paraffins and phosphate ester-plasticizers also act as flame retardants.

1.2.4 PLASTISOLS AND ORGANOSOLS

These are fluids in which fine PVC particles are dispersed in plasticizers. Plastisol or pastes (used synonymously) do not contain any solvent/volatile components. An organosol is a plastisol containing volatile organic solvents.

The viscosity of plastisols varies from pourable liquids to heavy pastes. PVC pastes have two important characteristics.

(1) They are liquids and can be processed in that condition. The processing conditions are determined by the property of the paste at ambient temperature.

(2) On application of heat, when required, they fuse to viscous solutions of polymer in plasticizer, and on cooling, they result in familiar plasticized PVC.

A typical formulation consists of resin, plasticizer, stabilizer, fillers, pigments, and viscosity modifiers. Unlike solid formulations, lubricants and polymeric modifiers are not added in pastes. The natures and roles of various ingredients are discussed below.

1.2.4.1 Resins

The requirements of paste polymers are rather conflicting. They should have the following:

a. Resistance at room temperature to the plasticizer for stability

b. Good affinity for plasticizer to rapidly dissolve in it at an appropriate temperature for proper gelation and fusion

Resins made by mass or suspension are porous granules and will absorb a high level of plasticizer to form a sticky agglomerate. They are not used for making plastisols. Paste resins are made by emulsion or microsuspension polymerization and are finished by spray-drying techniques. They have high sphericity and a fairly dense surface, so that penetration of plasticizer at room temperature is low. Particle size ranges from 0.1 μm to 3 μm. Particle size and particle size distribution profoundly influence the viscosity of the paste. Extender resins of particle size 80–140 μm having a nonabsorbent surface are generally added to lower the viscosity of the paste. Because emulsion-grade resins are expensive, incorporation of extender resins lowers the cost. Paste resins are usually homopolymers, but copolymers with vinyl acetate (3–10%) are used to lower the fusion temperature. Although paste resins are resistant to swelling or solution by plasticizer at ambient temperature, slow solvation still occurs, otherwise settling of the resin will take place. The slow solvation results in an increase in viscosity of the paste on storage, known as aging. Higher molecular weight resins give products with superior physical properties, but the fusion temperature is increased.

1.2.4.2 Plasticizers

The important factors in selecting a plasticizer are viscosity, viscosity stability, clarity, compatibility, permanence, and fusion temperature. A plasticizer should have adequate compatibility for fusion of the paste at an elevated temperature. Alkyl phthalates are the most common primary plasticizer.

1.2.4.3 Stabilizers

It is preferable to use liquid stabilizers for good dispersion. Care should be taken that the stabilizer is compatible with other liquids of the paste or else precipitation may occur. Ba-Zn and Ca-Zn combinations are generally used. Tin-based systems are used where clarity is desired.

1.2.4.4 Fillers

Various fillers like clay, calcium carbonate, barytes, etc., are added to the pastes. They affect the flow properties and aging characteristics of the paste. Fillers increase the paste viscosity due to an increase of the particulate phase and adsorption of plasticizer by the filler particles. The adsorption can be reduced by using coated fillers, such as with organic titanates.

1.2.4.5 Viscosity Depressants

These additives, which are surface-active agents, lower the viscosity and improve viscosity stability and air release properties. Polyethylene glycol derivatives are generally effective.

1.2.4.6 Thickeners

For certain applications, paste should have a high viscosity at low shear rates and a low viscosity at high shear rates. An example is in spray coating or dip coating where no sag/drip property is desired. Various thickening agents like fumed silica, special bentonites, and aluminium stearates are used. These form a gel structure, and the paste varies in consistency from butter to putty. They are also known as plastigels.

1.2.4.7 Blowing Agents

The addition of azo dicarbonamide, which decomposes to form nitrogen gas, is the common method used to produce expanded vinyl. The decomposition of the blowing agent should occur at or above the fusion temperature for the formation of a closed-cell structure. If the blowing agent decomposes completely before gelation, an open-cell structure will be formed. The cell size is determined by the rate of decomposition of the blowing agent and the melt viscosity of the fused composition. Low molecular weight polymers are used in foam formulation, so that the melt viscosity of the fused paste is low. Foam pastes are not commonly deaerated prior to use as the air present acts as nucleating agent for cell formation as the azo decomposes. Blow ratios are controlled by the quantity of azo compound. A high blow ratio will blow the foam apart during fusion.

1.2.4.8 Manufacture

The pastes are made in a simple paddle-type mixer that provides an intermediate level of shear. The temperature should not rise during mixing. The mixing is generally carried out in vacuum, or entrapped air is removed after mixing by subsequent deaeration. The presence of air may result in bubbles and loss of clarity of the end product.

1.2.4.9 Fusion

On heating, the liquid paste is converted into a solid. As the temperature of the paste rises, more plasticizer penetrates the polymer particles, causing them to swell. This process continues until at about 100°C, the liquid phase disappears completely with the formation of a gel (gelation temperature). On further heating, a solution of polymer and plasticizer is formed, with the formation of an homogenous plasticized PVC melt (fusion temperature). On cooling, solid plasticized PVC is obtained. The processes of gelation and fusion are the conversion of suspended polymer particles in a plasticizer to a solid containing dispersion of plasticizer in a continuous polymer matrix. This is, therefore, a phase inversion (Figure 1.5).

1.2.4.10 Organosols

Diluents are added to reduce the viscosity of plastisols, to make it suitable for spray, roller, brush, and other forms of coatings. The thinned plastisols are known as organosols. The diluents are nonsolvents of PVC, like toluene, xylene,

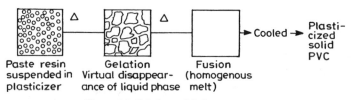

Figure 1.5 Gelation and fusion processes.

naphta, and mineral spirits. Addition of these diluents shifts the solubility parameter of the dispersing medium away from PVC, lowering the solvation and reducing the viscosity. As the diluent level increases, the viscosity passes through a minima and then increases with further dilution. This increase of viscosity is due to flocculation of the plastisol resin.

1.2.4.11 Uses in Coating

Pastes are extensively used for flexible coatings applied by dipping, spraying, or the spreading process. The products are diverse, including upholstery, luggage fabric, wall coverings, floor coverings, tarpaulins, and shoe uppers.

1.2.5 COMPOUNDING OF PVC FOR SOLID COMPOSITIONS

The purpose of compounding is to blend the resin and additives into a homogeneous, well-dispersed form appropriate for further processing. PVC formulations are used in the industry in liquid phase, i.e., paste and solutions, or in solid form, as powder or pellets. Solid phase compounding can be broadly categorized into two types, melt compounding and dry blending.

1.2.5.1 Melt Compounding

In this process, a premix is made in a ribbon or tumble blender. The premix is then fluxed and pelletized or may be directly sent to high shear mixers that break down the resin and simultaneously disperse and blend the additives with the fluxed resin. Two types of mixers are widely used: the batch-type mixers offer greater flexibility when frequent product and formulation changes are encountered and the continuous mixers are used when large volume and steady throughput is required. The batch-type mixers can be either a two roll mill or an internal mixer of the Banbury type, similar to those used for rubber compounding. Continuous mixers are single- or double-screw extruders.

1.2.5.2 Dry Blending

Dry blending is a process of adding liquid and dry compounding ingredients into PVC to produce a granular, free-flowing powder. The resin is not fused during the compounding operation. The individual particles of the dry blend are much like the initial resin. Their size is, however, greater due to absorption of plasticizer and other additives. The resins used for commercially plasticized applications are suspension and mass polymerized homopolymer of PVC. The resins are porous and have high surface area for rapid absorption of plasticizer, etc. It is desirable to select a resin with a narrow particle size range. The penetration of plasticizer in a large particle size causes "fish eyes" in the processed product. Fine particles tend to float in the molten mass causing surface imperfections in the product.

The dry blends are produced in different mixing equipment. Mainly, two types of mixers are used: the steam jacketed ribbon blender or the high-intensity batch mixer. They are coupled to a cooling blender. In a ribbon blender, the resin and dry ingredients are added first and allowed to mix for a short time to break the agglomerates. The blender may be heated, if required, to increase the absorptivity of the resin for the plasticizer. The heated plasticizer mix containing other liquid additives is then sprayed onto the resin mix, and the blender is heated. The time and temperature of the mix is dependent on the formulation, but normally the temperature is between 100°C to 130°C, and the time is between 10–20 min. The powder is next discharged to a coupled cooling mixer. In a high-intensity mixer, the procedure of adding the ingredients is similar. In these mixers, heating is mainly due to the mechanical energy of the mixing process, i.e., shear and friction.

Sintered dry blends are partly fluxed pellets that are obtained by heating the dry blend to near its fusion point to sinter the particles into agglomerates. These can vary in size from coarse powder to regular pellets. These sintered blends are nondusting, easy to transport, more homogenous than dry blends, and have better processability. The production is usually done in a high-intensity mixer coupled with a cooler mixer. After the dry blend is formed, the temperature of the blender is increased until the particles agglomerate to the desired size. The blend is then discharged rapidly into the cooling blender, where it is rapidly cooled, and further agglomeration is prevented.

The investment in equipment for dry blending is substantially low, enabling processors to carry out their own blending.

1.3 POLYURETHANES

1.3.1 INTRODUCTION

Polyurethanes [12–15] are polyaddition products of di- or polyisocyanate with a di- or polyfunctional alcohol (polyol).

$$n\,(\,NCO\text{-}R\text{-}NCO) + n\,(\,HO\text{-}R'\text{-}OH\,) \rightarrow$$

$$OCN\text{-}(\,\text{-}R\text{-}NH\text{-}\underset{\underset{O}{\|}}{C}\text{-}O\text{-}R'\text{-}O\text{-}\underset{\underset{O}{\|}}{C}\text{-}NH\text{-})_{n\text{-}1}\,\text{-}R\text{-}NH\text{-}\underset{\underset{O}{\|}}{C}\text{-}O\text{-}R'\text{-}OH$$

Urethane

(4)

If the functionalities of the reactants are three or more, branched or cross-linked polymers are formed. Variations in the R and R' segments of the polyaddition reaction shown above permit preparation of polyurethane to meet specific needs. The extent of cross-linking, chain flexibility, and intermolecular forces can be varied almost independently. The range of polyurethane products is thus quite diverse and includes fibers, soft and hard elastomers, and flexible and rigid foams.

The two important building blocks are isocyanates and polyols. Chain extenders like short-chain diols or diamines and catalysts are frequently used in the synthesis of the polymer. A brief account of the chemistry of the raw materials and polyurethanes is being discussed to obtain a proper perspective.

1.3.2 BUILDING BLOCKS OF POLYURETHANE

1.3.2.1 The Isocyanates

1.3.2.1.1 Basic Reactions of Isocyanates

The isocyanate group is highly reactive. Its most important reaction is the nucleophilic addition reaction of compounds containing an active hydrogen atom. The general equation is given by

$$\text{R-N=C=O} + \text{HX} \rightarrow \text{R-}\overset{\overset{\displaystyle H}{|}}{N}\text{-}\overset{\overset{\displaystyle O}{\|}}{C}\text{-X} \tag{5}$$

Some important reactions are given below:

(1) Reactions with compound containing -OH groups

$$\text{R-N=C=O} + \text{R' OH} \rightarrow \text{R-NH-}\overset{\overset{\displaystyle O}{\|}}{C}\text{-O R'} \tag{6}$$
$$\text{Urethane group}$$

The trivial name urethane, which is used for ethyl carbamate, is used as the generic name for all polyurethanes. Primary, secondary, and tertiary -OH show a decreasing order of reactivity.

(2) Reactions with compound containing -NH groups

$$R - N = C = O + R' \, NH_2 \rightarrow \quad R - NH - \overset{\displaystyle O}{\overset{\|}{C}} - NHR'$$

Substituted urea   (7)

Primary and secondary amines react vigorously forming substituted urea. Primary amines react faster than the secondary amines. Ammonia and hydrazines react similarly.

(3) The urethanes and ureas formed by the above reactions [Equations (6) and (7)] still possess acidic protons and react further with additional isocyanates to form allophanates [Equation (8)] and biurets [Equation (9)], respectively.

$$R - NH - \overset{\displaystyle O}{\overset{\|}{C}} - O \, R' + R - N{=}C{=}O \quad \rightarrow \quad R{-}N - \overset{\displaystyle O}{\overset{\|}{C}} - OR'$$

$$O = C - NH - R$$

Allophanate   (8)

$$R - NH - \overset{\displaystyle O}{\overset{\|}{C}} - NH{-}R' \; + R - N {=} C {=} O \rightarrow \quad R - N - \overset{\displaystyle O}{\overset{\|}{C}} -- NH{-}R'$$

$$O{=} C{-}NH{-}R$$

Biuret   (9)

In the case of polyisocyanates, the above reactions [Equations (8) and (9)] lead to the formation of branching in the polymer.

(4) Reaction with water: water reacts with isocyanate to form unstable carbamic acid that splits into carbon dioxide and the corresponding amine. The amine immediately reacts again with another molecule of isocyanate to form symmetrical urea. The carbon dioxide acts as a blowing agent in the production of foams.

$$R - N {=} C = O + H_2O \rightarrow \quad [\, R - NH - \overset{\displaystyle O}{\overset{\|}{C}} - OH] \; \overset{RNCO}{\rightarrow} \quad R - NH - \overset{\displaystyle O}{\overset{\|}{C}} - NH - R + CO_2$$

          (10)

(5) Amides react to form acyl urea

$$R - N = C = O + R' \, C \, O \, NH_2 \quad \rightarrow \quad R \text{-NH - } \overset{\overset{\displaystyle O}{\|}}{C} \text{ - NH - } \overset{\overset{\displaystyle O}{\|}}{C} \text{-R'}$$

Acyl urea

(11)

(6) Carboxylic acids: substituted amides are formed with liberation of carbon dioxide

$$R - N = C = O + R'COOH \dashrightarrow \quad [\, R \text{ - NH - } \overset{\overset{\displaystyle O}{\|}}{C} \text{-O -} \overset{\overset{\displaystyle O}{\|}}{C} \text{ - R'} \,] \quad \rightarrow \quad R\text{-NH-}\overset{\overset{\displaystyle O}{\|}}{C}\text{-R'} + CO_2$$

Substituted amide

(12)

(7) Self-addition reactions of isocyanate: highly reactive aryl isocyanates dimerize in the presence of catalysts to form uretidinediones. The formation of uretidinedione results in loss of reactivity of the isocyanate in storage. The uretidinedione formation is a means to block isocyanates and to make the isocyanate group available at elevated temperatures. Catalyzed by strong bases, isocyanates also undergo trimerization to form isocyanurate ring structure, which is very stable toward heat and most chemicals. In the case of polyisocyanate, highly branched polyisocyanurates are formed.

$$R - N = C = O + \; O = C = N - R \quad \rightarrow \quad \begin{array}{ccc} R\!-\!N & -\!\!\!- & C\!=\!O \\[-2pt] | & & | \\[-2pt] O\!=\!C & -\!\!\!- & N\!-\!R \end{array}$$

Uretidinedione

(13)

1.3.2.1.2 Important Polyisocyanates and Their Synthesis

The most important route for the synthesis of isocyanates is the phosgenation of primary amines.

$$R\text{-}NH_2 + COCl_2 \rightarrow R\text{-}N{=}C{=}O + 2\,HCl \tag{14}$$

Toluene diisocyanate (TDI) is one of the most important diisocyanates. The synthetic steps are dinitration of toluene, reduction to the corresponding diamines, followed by phosgenation to yield TDI. It is used as a mixture of 2,4 and 2,6 isomers in the ratio of 80:20. The two isomers differ considerably in reactivity, so the actual ratio of the two components is quite important.

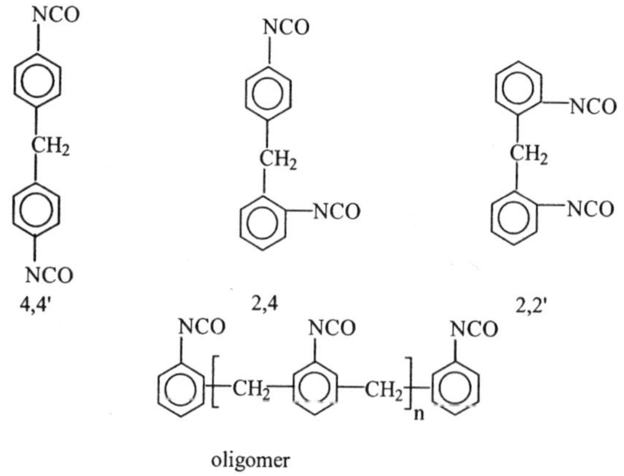

Figure 1.6 Isomers of MDI.

Another important aromatic diisocyanate is diphenyl methane diisocyanate (MDI). Condensation of aniline with formaldehyde leads to a mixture of 4,4'-, 2,2'- and 2,4'-diamino diphenyl methanes as well as polyamines. These on phosgenation form the corresponding isocyanates (Figure 1.6).

Other common aromatic isocyanates used are naphthalene 1,5-diisocyanate, xylelene diisocyanate XDI, p-phenylene diisocyanate PPDI, and 3,3'-tolidine diisocyanate TODI.

Aromatic isocyanates yield polyurethanes that turn yellow with exposure to light. Various aliphatic and cycloaliphatic diisocyanates are used in the industry to produce polyurethanes, which do not turn yellow upon light exposure. These are extensively used for coatings. The most important among the aliphatics is hexamethylene diisocyanate (HMDI) obtained by the reduction of adiponitrile and phosgenation of hexamethylene diamine. Among the cycloaliphatics, isophorone diisocyanate (IPDI) is the most common. Others are cyclohexyl diisocyanate (CHDI), 4,4'-dicyclohexyl methane diisocyanate (H_{12}MDI), and 2,2,4-trimethyl-1,6-hexamethylene diisocyanate (TMDI).

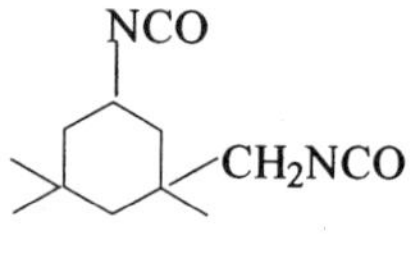

1.3.2.1.3 Blocked Isocyanates

In blocked isocyanates, the isocyanate groups are reacted with compounds to form a thermally weak bond. On heating, the bond dissociates to regenerate the isocyanate group. The most common example is the reaction product of phenols to yield aryl urethanes that dissociate at ~150°C.

$$R\text{-}N{=}C{=}O + ArOH \rightarrow R\text{-}NH\text{-}COOAr \qquad (15)$$

Similar adducts are formed by the reaction of isocyanates with diphenylamine, succinimide, acetoacetic ester, oximes, triazoles, caprolactams, etc. Dimers of isocyanates can also be considered as blocked isocyanate. The ring opening can be thermal or catalytic, without the liberation of any volatile blocking agent.

2,4TDI

$$(16)$$

The blocked isocyanates are specifically used in one component systems for coatings adhesives. They are also used for preparing aqueous polyurethane dispersions. The blocked isocyanate group does not react with water. The isocyanate group is regenerated for reaction after drying and heating of the dispersion.

1.3.2.2 Polyols

Besides the polyisocyanates, the other important building blocks of polyurethanes are polyfunctional alcohols. Polyurethanes made from short-chain diols yield linear crystalline fiber-forming polymers, lower melting than the corresponding polyamides, that are of little commercial interest. However, reaction of isocyanates with polymeric glycols lead to the formation of polyurethanes of diverse physical and mechanical properties, suitable for a variety of applications. The molecular weight of the polyols ranges from 200 to 10,000. Two types of polyols are used in the synthesis of polyurethane: polyester polyols and polyether polyols containing at least two hydroxyl groups in a polyester or polyether chain.

1.3.2.2.1 Polyester Polyols

These are saturated dicarboxylic esters, the reaction product of dibasic acid and a diol. The commonly used acids are adipic, phthalic, sebacic, and dimer acids (dimerized lineoleic acid). The diols are ethylene glycol, diethylene glycol, triethylene glycol, 1,2-propylene glycol, etc. For higher functionality, glycerol, trimethylol propane, penta erythritol, sorbitol, etc, are used where chain branching and higher cross-linking are required.

For preparation of the polyester, conventional methods of polyesterification are used. The molecular weight can be controlled by the molar ratio of the reactants and the reaction conditions, however, it is essential that the terminal groups be hydroxyl for reaction with the isocyanate. For this purpose, esterification is carried out with stoichiometric excess of the diol.

$$H\text{-[OR-OOCR'-CO---]}_n\text{-OROH}$$

Polyester polyol

Caprolactone polyester formed by polyaddition with a diol and ε-caprolactone in the presence of an initiator is also of commercial interest. The advantage of this reaction is that no water is formed.

$$n \quad \text{(caprolactone ring)} \quad \begin{array}{c} HO\text{-R-OH} \\ \rightarrow \end{array} \quad HO\text{-R-O-[C-(CH}_2)_5\text{-O-]}_n\text{- H}$$

Caprolactone polyester   (17)

1.3.2.2.2 Polyether Polyols

Polyether polyols are also known as polyalkylene glycols or polyalkylene oxides. The common polyether polyols are polypropylene glycols and polytetramethylene glycol (Figure 1.7).

Polypropylene glycol, the most commercially important polyol, is prepared by a base-catalyzed ring opening polyaddition reaction. However, polytetramethylene glycol is obtained by acid-catalyzed ring opening polyaddition.

Block copolymers of polyethylene and polypropylene glycols are commonly used in the industry. These are obtained by reacting ethylene oxide with polypropylene glycol or propylene oxide with polyethylene oxide in the presence of a base catalyst. This offers a means of adjusting the ratio of primary and secondary hydroxyl groups. The general formula of the block copolymer is given by the following:

$$HO\text{-(CH}_2\text{-CH}_2\text{-O-)}_a\text{- (CH}_2\text{-CH-O-)}_b\text{-(CH}_2\text{-CH}_2\text{-O-)}_c\text{-H}$$
$$\overset{\displaystyle CH_3}{\vert}$$

Copolymer of ethylene and propylene oxide

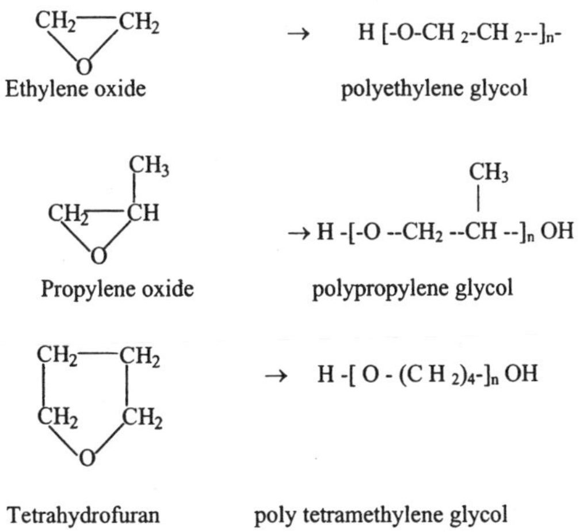

CH$_2$——CH$_2$ →   H [-O-CH$_2$-CH$_2$--]$_n$-
    \ /
     O
Ethylene oxide   polyethylene glycol

                CH$_3$ CH$_3$
                | |
CH$_2$——CH   → H -[-O --CH$_2$ --CH --]$_n$ OH
    \ /
     O
Propylene oxide   polypropylene glycol

CH$_2$——CH$_2$ →   H -[O - (C H$_2$)$_4$-]$_n$ OH
| |
CH$_2$ CH$_2$
    \ /
     O

Tetrahydrofuran   poly tetramethylene glycol

Figure 1.7 Common polyether polyols.

1.3.2.3 Cross-linkers and Chain Extenders

These are low molecular weight polyfunctional alcohols and amines that act as chain extenders or cross-linkers by reaction with the -NCO group. The alcohols form urethane, and the amines form urea linkages. The difunctional compounds are essentially chain extenders, while the compounds with functionality greater than two are cross-linkers. The end properties of the polyurethanes are considerably influenced by these compounds, as they alter the hard to soft segments proportion of the polymer. Some important diol chain extenders are ethylene glycol, 1,4-butanediol, etc. Trifunctional alcohols like glycerol and trimethylol propane act as cross-linkers. Among the amines, derivatives of di-amino phenyl methane and *m*-phenylene diamines are of commercial interest as chain extenders. The most common in this category is 3,3′dichloro 4,4′diamino diphenyl methane (MOCA).

1.3.2.4 Catalysts

The rate of the reactions of isocyanates can be greatly enhanced by using appropriate catalysts. The most important catalysts are the tertiary amines and organo tin compounds. The rate increase of urethane link formation by -NCO/OH reaction depends on the basicity and the structure of the amines. Some important amines are triethyl amine, triethylene diamine and peralkylated aliphatic polyamines. Prominent among the organo tin compounds are dibutyl tin dilaurate and tin dioctoate. They are readily soluble in the reaction mixture and have low volatility and little odor. Basicity favors formation of branching and cross-linking. Commercially, a mixture of amine and tin catalysts are used for synergistic effect.

1.3.3 METHOD OF PREPARATION OF POLYURETHANES

There are two methods of preparation of polyurethanes.

(1) *One-shot process:* in this process, the entire polymer formation takes place in one step by simultaneously mixing polyol, diisocyanate, chain extender, and catalyst. The reaction is very exothermic and requires similar reactivities of different hydroxy compounds with the isocyanate.

(2) *Prepolymer process:* this is a two-stage process. In the first stage, diisocyanate and polyol are reacted together to form an intermediate polymer of a molecular weight of about 20,000, which is called a prepolymer. Depending on the stoichiometry of the diisocyanate and polyol, the prepolymer can be NCO terminated or OH terminated. The NCO-terminated prepolymers are of great technical importance, as the NCO groups are available for reaction with compounds containing active hydrogen atoms. The prepolymer is then reacted with a chain extender to form the final high molecular weight polymer, either by a polyfunctional alcohol or amine (Figure 1.8).

1.3.4 STRUCTURE OF POLYURETHANES

Polyurethanes prepared from short-chain diols and diisocyanate have a large concentration of urethane linkage which results in a high degree of hydrogen bonding between the -NH and C=O groups of the chains. Consequently, these polymers are hard and have a low degree of solubility. On the other hand, the reaction product of long-chain polyols and diisocyanates results in polymers with a low concentration of urethane groups. The intermolecular forces are, therefore, mainly weak van der Waals forces, and the polyurethane is low in hardness and strength. Most polyurethanes are prepared from at least three basic starting materials, viz., (1) long-chain polyol, (2) diisocyanate, and (3) a chain extender. These linear chains of polyurethane elastomer show segmented structure (block copolymer), comprised of an alternate soft segment of the polyol with weak interchain interaction, present in coiled form, and a hard segment formed by reaction of diol/diamine and the diisocyanate (Figure 1.9). The hard segments have strong interchain H-bonding and dipolar interactions due to the presence of a large number of polar groups—urethanes and ureas.

The hard and soft segments are partly incompatible because of their difference in polarity; as such, they show two-phase morphology. The hard segments

OH~~OH + OCN-R'-NCO → OCN ~~~~~NCO → chain extension by
  polyol   diisocyanate   NCO terminated prepolymer

diol/diamine   →   Final polyurethane

Figure 1.8 Prepolymer process.

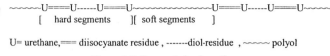

[hard segments][soft segments]

U= urethane,=== diisocyanate residue , -------diol-residue , ~~~~ polyol

Figure 1.9 Segmented polyurethane.

form discrete domains in a matrix of soft segments. The aggregated hard segments tie the polymeric chains at localized points, acting as cross-links and as reinforcing filler matrix in a soft segment matrix (Figure 1.10). The two-phase morphology is applicable to linear elastomers as well as to most of the cross-linked polyurethanes.

1.3.5 THERMOPLASTIC URETHANE ELASTOMERS [13,16]

TPUs are high molecular weight polymers obtained by the reaction of polyol, diisocyanate, and chain extender. These are fully reacted linear chains, segmented in structure with hetero-phase morphology as described in the previous section. If however, stoichiometric excess of isocyanate is taken, i.e., NCO/OH > 1.0, the free NCO groups react with urethane groups forming allophanate branched or cross-linked structures.

The cross-links formed due to hard segment domains impart elasticity to the polymer, however, these are reversible to heat and solvation, permitting thermoforming and solution application of TPU. They can, therefore, be termed virtual cross-links. The soft matrix has T_g lower than room temperature, and is amorphous in nature, but the hard segments are paracrystalline or crystalline.

The property of the elastomer is dependent on the type of polyol, molecular weight, and the ratio of hard to soft segments. The molecular weight of the polymer for optimum physical properties is between $\bar{M}_w$ 100,000–200,000. Polyethers generally give elastomers having a lower level of physical properties than the polyester polyols. The elastomer can be compounded in a plastic or rubber mill. Stabilizers, processing aids, and extenders are the additives used. They do not require any curing.

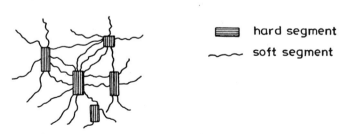

▬ hard segment

～ soft segment

Figure 1.10 Two-phase morphology of polyurethane.

1.3.6 POLYURETHANE COATINGS [12,14,17]

The conventional solution-based coatings are of two types, viz., one-component systems and two-component systems.

1.3.6.1 One-Component System

These are two types of one-component systems: reactive and completely reacted systems.

a. Reactive one-component systems: these systems are low molecular weight prepolymers with terminal isocyanate groups. They are dissolved in solvents of low polarity. After coating, they are moisture cured. The water acts as a chain extender and cross-linking agent with the formation of urea and biuret linkages. The generation of carbon dioxide is sufficiently slow, so that slow diffusion of the gas from the film occurs without bubble formation. The rate of cure is dependent on the temperature of the cure and the humidity of the ambient. Use of blocked isocyanate prepolymers allows formulations of one-component systems that are stable at room temperature.

b. Completely reacted one-component system: this consists of totally reacted high molecular weight thermoplastic polyurethane elastomers. The PU is dissolved in a highly polar solvent like dimethyl formamide. These coatings dry physically.

1.3.6.2 Two-Component System

In this type of coating system, isocyanate-terminated prepolymers or polyfunctional isocyanates are reacted with polyhydroxy compounds that may already be urethane modified. The polyisocyanate component, usually in the form of a solution, is mixed with the polyhydroxy component prior to coating. Curing of these coatings occurs due to the formation of urethane linkages. In addition, reaction with moisture also takes place. The properties of the resulting coatings depend on various factors, viz.,

(1) The polyol type and molecular weight
(2) The temperature of the reaction
(3) The concentration of polar groups, i.e., urethane and urea
(4) The cross-linking density

In the U.S., for PU-coated fabrics, TPU systems are preferred. By varying the polyol and NCO/OH ratio in the TPU manufacturing process, the same wide range of flexibility can be obtained as mentioned for the two-component systems. Proper choice of TPU adhesive is critical. It is common practice to add a polymeric isocyanate to the TPU adhesive layer to aid in adhesion to the base fabric. There is almost no transfer coating done in the U.S. today for wearing apparel. It is all done overseas. In the U.S., fabrics for tenting, recreational clothing, etc., are direct coated with TPU solvent system.

Solvent-free coating by TPU elastomers: these can be coated on the fabric by hot melt process of the solid polymer. The common method employed is the Zimmer coating method. The PU can also be extrusion fed to a Bema or a calender.

1.3.6.3 Additives

The additives used for urethane coatings are generally silica fillers to reduce gloss, UV absorbers, antioxidants, and flow improvers. The solvents used for coating should be free of moisture and reactive hydrogen to prevent reaction with free isocyanate in two-component systems. They are generally polar in nature, and care should be taken for their selection to ensure storage stability and blister-free film. Similarly, the pigment used should also be moisture free.

1.3.7 AQUEOUS DISPERSION OF POLYURETHANES [12,13,18–21]

In recent years, there has been a trend to use PU latices for coating for the following reasons.

- low emission of organic volatiles to meet emission control regulations
- lower toxicity and fire hazard
- economy of the solvent
- viscosity of latex independent of molecular weight

Chemically PU latices are polyurethane-urea elastomers dispersed in water. The starting materials are polyether/polyester polyol, diisocyanates, and polyfunctional amines-chain extenders. Isocyanates should have low reactivity to water, as the carbon dioxide produced leads to foaming. For chain extension purposes aliphatic or cycloaliphatic amines are preferred, because they react faster with the isocyanate group.

1.3.7.1 Emulsifiers

Hydrophobic polyurethane can be dispersed in water with a protective colloid. Alternately, hydrophilic groups can be incorporated in the polymer chain by internal emulsifiers that have the following advantages:

a. Dispersion does not require strong shear force

b. Leads to better dispersion stability

c. Finer particle size

These emulsifiers are of two types: ionic and nonionic. Ionic internal emulsifiers are anionic or cationic groups built in the polymer chain. Common ionic emulsifiers are as follows:

(1) Sulfonated diamine: $H_2N\text{-}CH_2\text{-}CH_2\text{-}NH\text{-}CH_2\text{-}CH_2\text{-}SO_3^-\ Na^+$

(2) Sulfonated diols: $HO\text{-}CH_2\text{-}CH_2\text{-}\underset{\underset{SO_3^-\ Na^+}{|}}{C}H\text{-}CH_2\text{-}OH$

(3) Dihydroxy carboxylic acids: $HO\text{-}CH_2\ \underset{\underset{COOH}{|}}{\overset{\overset{CH_3}{|}}{C}}\ CH_2\ OH$

(4) A tertiary amine: $HO\text{-}CH_2\text{-}CH_2\text{-}\underset{\underset{R}{|}}{N}\text{-}CH_2\text{-}CH_2\text{-}OH$

Polyurethane ionomers with built in ionic/hydrophilic groups are obtained by reacting NCO terminated prepolymer with ionic internal emulsifiers. Nonionic internal emulsifiers are polyether chains of polyethylene oxide, which are incorporated in the PU chain. These segments, being hydrophilic in nature, act as emulsifiers of the elastomer. The disadvantage of such emulsification is the water sensitivity of the dried film. Anionic dispersions are more widely used.

Typically, polymer particle size ranges from 0.01 to 0.1 micron. In the ionically stabilized dispersions, the ionic centers are located at the surface, and the hydrophobic chain segments are at the interior. They are sensitive to electrolytes. Nonionically stabilized dispersons are sensitive to heat, as polyethylene glycol polyol lose their hydrophilicity at higher temperatures.

1.3.7.2 Preparation of Dispersions

There are different methods for preparing dispersions [12,18], e.g., acetone process melt dispersion process, etc.

1.3.7.2.1 Acetone Process

A solution of high molecular weight polyurethane-urea ionomer is built up (after reaction and chain extension) in a hydrophilic solvent like acetone, dioxane, or tetrahydrofuran. The solution is then mixed into water. On removal of the solvent by distillation, an aqueous dispersion of the polymer is obtained

OCN~~~~~NCO + H₂ N CH ₂-CH ₂-NH   +   OCN~~~~~NCO

$$|$$

CH ₂-CH₂ -SO₃⁻ Na⁺

~~~~~NH -C -NH -CH₂ -CH ₂-N-C-NH ~~~~

    ‖ | ‖

    O | O

                         |

              CH₂-CH₂-SO₃⁻ Na⁺

↓water

dispersion of PU in acetone + water

↓solvent removal

aqueous dispersion of polyurethane urea

Figure 1.11 Acetone process of PU dispersion.

by phase inversion of the emulsion originally formed with the water-organic solvent. A typical reaction sequence is shown above (Figure 1.11).

1.3.7.2.2 Melt Dispersion Process

A -NCO terminated prepolymer is reacted with ammonia or urea to form urea or biuret end groups. The reaction with urea is carried out at a high temperature ∼130°C. The hot melt is poured into water at an elevated temperature to form a spontaneous dispersion. The end capping of -NCO groups renders them nonreactive to water. Chain extension is carried out by reaction of the oligomer with formaldehyde through the formation of methylol groups at the biuret functionality at a lower pH. The reaction sequence is shown in Figure 1.12.

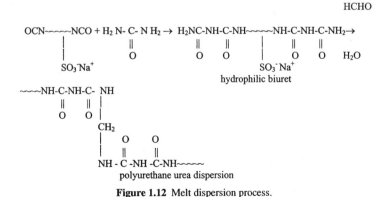

Figure 1.12 Melt dispersion process.

1.3.7.3 Film Properties [18,21]

On drying of the dispersion on a substrate, the discrete polymer particles should fuse to form a continuous organic phase with entanglement of polymer chains. Poor fusion leads to poor gloss and poor physical properties of the film. If cross-links are present, the film-forming property decreases. An improvement can be made by adding high boiling, water miscible solvent in the latex. On evaporation of water, a solution of PU in the solvent is left behind. The solvent on evaporation gives a continuous film. A commonly used solvent is *N*-methyl pyrrolidone.

The main drawbacks of PU dispersion compared to two-component solution are the poor solvency and water resistance. Improvement in these properties can be obtained two ways:

(1) Grafting hydrophobic chains, usually acrylics, on the PU backbone
(2) Cross-linking of the polyurethane chains particularly those containing carboxylate ion, using polyfunctional aziridines

For coating purposes, certain additives are added in dispersion. These include thickening agents (e.g., polyacrylate resins), extenders, pigments, flame retardants, and external cross-linking, e.g., aziridines and melamine resins.

Great improvements have been made in water-based polyurethane chemistry; however, the wide flexibility of properties that can be obtained by solvent-based systems has not been demonstrated with the water-based coatings. To date, where strength, flexibility, toughness, etc., are the required physical properties of the end product, solvent-based systems are the coatings of choice.

1.3.8 FEATURES OF POLYURETHANE COATING

Polyurethane coating on textiles gives a wide range of properties to the fabric to meet diverse end uses like apparel, artificial leather, fuel and water storage tanks, inflatable rafts, containment liners, etc. This is because of a wide selection of different raw materials for their synthesis. For a breathable microporous coating, PU is the polymer of choice. PU-coated fabric offers advantages, which are given below, over other polymeric coatings [20,22,23].

- dry cleanability, as no plasticizers are used
- low temperature flexibility
- overall toughness—very high tensile, tear strength, and abrasion resistance requiring much less coating weight
- softer handle
- can be coated to give leather-like property and appearance

1.3.8.1 Structure-Property Relationship [13,24–26]

As has already been discussed, polyurethanes show two-phase morphology of soft and hard segments. The types and content of these segments profoundly influence the properties of the polyurethane. The soft segments obtained from the polyol determine the elastic and low temperature properties. Increasing the molecular weight of the polymer by increasing the soft segments results in lowering of the mechanical property, making the polymer softer and more extensible. Similarly, the presence of pendant side groups in the polyol also results in lowering of the physical properties. Increase in hard segment in the polymer increases its hardness, tensile modulus, and tear resistance. Diamine chain extenders form urea linkages, that enter into stronger hydrogen bonds and yield stiffer hard segments, thus, the PU formed has higher hardness and modulus than that obtained by diol extenders.

Polyester polyurethanes generally show higher modulus, tensile strength, hardness, and thermal oxidative stability than the polyether urethanes. This is because of the higher cohesive energy of the polyester chains. They also show better resistance to hydrocarbons, oils, and greases, but they show poorer hydrolytic stability due to the ester linkages. The hydrolytic stability of the polyester polyurethane can be increased by using sterically hindered glycols like neopentyl glycol and long-chain or aromatic diacid-like terephthalic acid.

Hydrolytic stability increases with hydrophobicity of the chain, thus, polyether polyurethanes have better hydrolytic stability. The thermal oxidative stability of polyether polyurethanes can be improved by adding antioxidants. Polyether urethanes also show better resistance to mildew attack.

The structure of isocyanates also influences the properties of the polyurethane. Symmetrical aromatic diisocyanates like naphthalene diisocyanate, diphenyl methane diisocyanate, and *p*-phenylene diisocyanate give a harder polymer with a higher modulus and tear strength compared to those obtained using less symmetrical ones such as 2,4- and 2,6-TDI. Polyurethanes of aliphatic and cycloaliphatic diisocyanates are less reactive and yield PU that have greater resistance to UV degradation, and thermal decomposition. They do not yellow on weathering but have a lower resistance to oxidation. The yellowing of polyurethanes containing aromatic diisocyanates is due to UV-induced oxidation that results in the formation of quinone imide [27] (Figure 1.13).

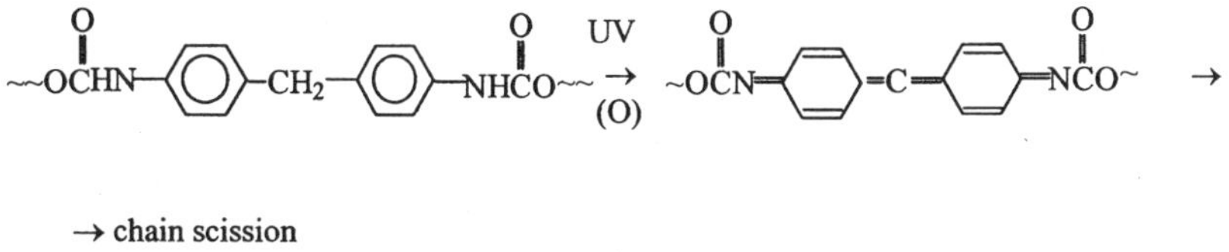

→ chain scission

Figure 1.13 UV-induced oxidation of MDI.

1.4 ACRYLIC POLYMERS

They are commonly known as acrylics. The monomers are esters of acrylic and methacrylic acid.

$$H_2C=C\overset{R}{\underset{\underset{O}{\overset{\|}{C}}-OR'}{}}$$

Acrylic ester

This is the general formula of acrylates (R = H for acrylates, R = CH$_3$ for methacrylates). Some common esters are methyl, ethyl, n-butyl, isobutyl, 2-ethyl hexyl, and octyl. The esters can contain functional groups such as hydroxyl, amino, and amido. The monomers can be multifunctional as well, such as trimethylol propane triacrylate or butylene glycol diacrylate. The nature of the R and R' groups determines the properties of monomers and their polymers. Polymers of this class are noted for their outstanding clarity and stability of their properties upon aging under severe service conditions.

Polymerization of the monomers occurs by free radical polymerization using free radical initiators, such as azo compounds or peroxides. Acrylic polymers tend to be soft and tacky, while the methacrylate polymers are hard and brittle. A proper adjustment of the amount of each type of monomer yields polymers of desirable hardness or flexibility. A vast majority of commercially available acrylic polymers are copolymers of acrylic and methacrylic esters. The polymerization can occur by bulk, solution, emulsion, and suspension methods. The suspension-grade polymer is used for molding powders. The emulsion and solution grades are used for coatings and adhesives.

Acrylate emulsions are extensively used as thickeners and for coatings. Acrylics have exceptional resistance to UV light, heat, ozone, chemicals, water, stiffening on aging, and dry-cleaning solvents. As such, acrylics are used as backcoating materials in automotive upholstery fabric and carpets, window drapes, and pile fabrics used for outerwear.

1.5 ADHESIVE TREATMENT

1.5.1 MECHANISM OF ADHESION

The adhesion of the polymer to the textile substrate is an important aspect of coating technology, especially when the articles are put to dynamic use. The main mechanisms of adhesion are as follows [28]:

(1) Mechanical interlocking: this mechanism operates when the adhesive interlocks around the irregularities or pores of the substrate, forming a mechanical anchor. A rough surface has a higher bonding area.

(2) Adsorption: the attractive forces may be physical, i.e., physical adsorption by van der Waals forces, H-bonding, or chemical bonding (chemisorption).

(3) Diffusion: the adhesive macromolecules diffuse into the substrate, i.e., interpenetration occurs at the molecular level. It requires that the macromolecules of the adhesive and the adherend have sufficient chain mobility and are mutually compatible.

The irregularities on the textile substrate for mechanical interlocking of the elastomer are fiber ends, twists, crimps of the yarn, and interstices of the weave pattern. Cotton fabric and yarns made from staple fiber have a much higher surface area, and the fiber ends become embedded in the elastomeric matrix. In contrast, the synthetic fibers normally produced in continuous filament are smooth and, hence, have relatively poor adhesion.

All textiles of practical interest have surfaces that contain oriented dipoles that induce dipoles in the elastomer. As such, the contribution of dipole-induced interactions is quite prominent. Hydrogen bonding also provides significant contribution to many coatings. Chemical bonding due to formation of crosslinks is also important, particularly when keying agents or special adhesive treaments are given to the fabric.

Textile substrates made from cotton or a high proportion of cotton do not generally require an adhesive treatment because the mechanical interlocking of the staple fiber ends into the elastomeric matrix imparts adequate adhesion. Rayon, nylon, and polyester used mainly in continuous filament forms require an adhesive pretreatment. The type of bonding systems in use are the resorcinol-formaldehyde-latex (RFL) dip systems, dry bonding systems and isocyanate bonding.

1.5.2 RESORCINOL-FORMALDEHYDE-LATEX SYSTEM [3,29,30]

This system is used for adhesive treatment of rayon and nylon. The latices used are generally of natural rubber, SBR, or vinyl pyridine copolymer (VP). A typical method of preparation consists of mixing resorcinol and formaldehyde in the molar ratio of 1:1.5 to 1:2.5 (in an alkaline medium resole process) in a rubber processer. The mixture is stored for about 6 hrs at room temperature. To this is then added appropriate latex or latex mixture. This mix is matured for 12–24 hrs prior to use. Many factors affect the performance of the RFL dip such as (1) the resorcinol-to-formaldehyde ratio (2) the pH and conditions of reactions, (3) the type of latex used, (4) the ratio of latex to resin, and (5) total solids, etc. A SBR-VP latex mixture is commonly used for rayon, and nylon. For standard tenacity rayon, the ratio of SBR to VP latex is about 80:20, but for higher tenacity yarns, a higher ratio (50%–80%) of VP latex is used. The solid content of the dip is adjusted to the type of fiber used, e.g., about 10–15 wt.% for rayon and about 20 wt.% for nylon. For nylon 6 and 66, better adhesion is achieved at 75% VP latex or higher.

The process of application consists of impregnating the fabric by passing through a bath of the dip. Excess dip is removed by squeezing through rollers. The water is then removed by drying at around 100–120°C. The treated fabric is then baked by heating at temperatures ranging from 140–160°C for a short period of time (1–2 mins). Total solid pickup is controlled mainly by solid content of the dip. The add-on required depends on the fiber. For rayon, the add-on ranges from 5–8 wt.%, for heavier nylon fabrics, an add-on of up to 15% may be required. In the case of nylon, curing is combined with heat setting at 170–200°C. The baking enhances adhesion due to increased condensation of the resin and creation of more reactive sites [31].

The RFL dip discussed above is adequate for olefin rubbers. For polymers like CR, NBR, PVC, IIR (butyl), and EPDM, the RFL dip has to be modified for better adhesion. With CR and NBR, replacement of latex of the RFL dip by 50–100% of the latex of the corresponding polymer gives satisfactory results. Latices of IIR and EPDM do not exist, although their emulsions are available that can be used to substitute the latex of RFL dip. However, the performance with emulsions is not satisfactory. For PVC, the latex has to be carefully selected, as all PVC latices do not form coherent film, and it may be desirable to incorporate an emulsion of the plasticizer in the dip for proper film formation.

Polyester fabric requires a two-stage dip. In stage one, the fabric is dipped into an adhesive consisting of water-miscible epoxy (derived from epichlorohydrin and glycerol) and a blocked isocyanate dispersion, to give a pickup of about 0.5%. The blocked isocyanate is activated at about 230°C. In stage two, the fabric is dipped in a standard RFL system. A two-stage dip is also used for aramid fabrics.

Estimation of add-on of RFL dip is critical for determining the adhesion level. The conventional ASTM method based on acid extraction of the coated material is time consuming. Faster results have been reported using near infrared (NIR) reflectance measurement [32].

1.5.3 DRY BONDING SYSTEM [3,30,33]

In this system, the adhesion promoting additives are added to the elastomer compound itself. Adequate bonding is achieved by coating the compound on untreated textile substrate. The important ingredients of the system are resorcinol, a formaldehyde donor-hexamethylene tetramine (HMT), and hydrated silica of fine particle size. The resorcinol formaldehyde resin formation takes place during the vulcanization process, which then migrates to the rubber-textile inerface, resulting in an efficient bond between the two surfaces. The role of silica is not fully clear. It probably retards the vulcanization process, allowing time for the formaldehyde donor to react with resorcinol and the resin to migrate to the interface for bond formation. The process is universally applicable to all types of rubber in combination with all types of textile materials. With polyester, however, hexamethoxymethyl melamine (HMMM) is used as a methylene donor instead of HMT, as the amine residue of HMT degrades the polyester by ammonolysis of the ester linkage. Because this system acts by migration of the resin components, a minimum thickness of the adhesive compound is required at the interface to prevent back migration of the adhesive components to the bulk. The normal amounts of resorcinol and HMT added are around 2.5 phr and 1.5 phr, respectively, however, the concentration is dependent on the formulation of the compound and the type of fabric. For dry bonding, the composition of the rubber compound should be carefully balanced. Nonsulfur curing systems lead to poorer adhesion. Use of ultraaccelerators is unfavorable because they often do not give sufficient time for the release of the required amount of formaldehyde for resin formation. On the other hand, too much delay in curing also leads to poor adhesion. Accelerators like *N*-cyclohexyl-2-benzothiazyl sulfonamide (CBS), used alone or in combination with basic secondary accelerators like *N,N*-diphenyl guanidine (DPG), give good adhesion.

1.5.4 MODE OF ACTION OF RFL DIP

The mechanism of action of the RFL system has been investigated. The RFL film forms a bond with the coated elastomer as well as the textile substrate. The dip film to elastomer bond occurs due to cross-linking of the rubber (latex) part of the film by migration of sulfur and curatives from the main elastomer. A minor contribution is due to the reaction of the resin part of the dip film with the active hydrogen atom of the main elastomer, forming a chroman-like structure, as given below [30] [Equation (18)]. Such a mechanism and formation of an interpenetrating network explains the adhesion of the dry bonding system.

Chroman-like structure

(18)

The adhesion of the textile substrate with the dip is believed to be due to its resin component. Various mechanisms are operative; they are mechanical interlocking, diffusion, and chemical bond formation. The methylol group of resins reacts with the hydroxyl groups of the cellulosics and amido group of nylon forming covalent bonds. This mechanism is applicable to the dry bonding system as well. In the case of polyester, the blocked isocyanate of the two-stage system forms a polyurethane with a solubility parameter similar to that of polyester, thus favoring adhesive bond by diffusion. The new surface is reactive toward the second-stage RFL dip [30].

1.5.5 ISOCYANATE BONDING SYSTEM

Polyisocyanates are used for binding elastomer and fabric. The isocyanate groups are highly reactive, and they bond elastomer and textile by reacting with their reactive groups. The common isocyanates used are 4,4,4-triphenyl methane triisocyanate (TTI), diphenyl methane diisocyanate (MDI), dianisidine diisocyanate (DADI), and polymethylene polyphenyl isocyanate (PAPI). There are two general procedures, viz., the solution process and the dough process [13]. In the solution process, a dilute solution of the polyisocyanate (~2% concentration) in toluene or methylene dichloride is applied on the fabric by spraying or dipping. After evaporation of the solvent, an elastomeric coating is applied in the usual manner. In the dough process, compounded rubber stock is dissolved in a suitable solvent like toluene, chlorobenzene, or gasoline, and is then mixed in a sigma mixer. To this solution (cement) is added isocyanate solution with agitation. This solution is then applied on the fabric using a conventional method to an add-on of about 10–15% and then dried. This coating acts as a primer for adhesion of the elastomeric coat to the fabric. The treated fabric in this case is better protected from moisture than in the solvent process. The composite-coated fabric can be cured in the usual manner. Isocyanate bonding agents give coated fabrics better softness and adhesion than that provided by the RFL system. The use of blocked isocyanate along with the rubber dough increases the pot life of the adhesive dip.

The adhesive treatments discussed in this section are primarily for elastomeric coatings on fabric. The formulation systems for other polymers and the factors responsible for proper adhesion have been discussed in Chapter 4.

1.6 RADIATION-CURED COATINGS

1.6.1 GENERAL FEATURES

Conventionally, curing of a polymer composition is done by thermal energy from sources such as electrical heaters, high pressure steam, hot air from electric heaters and infrared heaters. Curing by heat generated by a microwave is also being used for continuous vulcanization of rubber compounds. Radiation curing, i.e., curing by ultraviolet and electron beam radiation, does not involve heating, instead ions or free radicals are generated and the macroradicals so generated couple together to produce a three-dimensional network. In both UV/E beam, the formulation used for coating consists of an oligomer, a reactive monomer, and in the case of UV curing, a photoinitiator. Radiation-cured coatings have several advantages [34–36].

- fast curing speeds
- high solid content—usually 100% solids
- compact curing lines and decreased floor space
- low capital investment
- ability to cure heat-sensitive substrates
- wide variety of formulations
- reduced pollution
- lower energy consumption

1.6.2 CHEMISTRY

The chemistry discussed here is for ultraviolet radiation technology. In a general sense, the same formulations are used for both UV and E beam, except that in the case of UV curing, a photoinitiator is required to initiate free radical formation. The formulation of UV curing consists of the following [35–37]:

(1) A monomer and/or an oligomer bearing multifunctional unsaturated groups

(2) A photoinitiator that must effectively absorb incident UV light and produce initiating species with high efficiency

1.6.2.1 Photoinitiators

The different photoinitiators can be classified into three major categories:

(1) Free radical formation by homolytic cleavage: Benzoin alkyl ethers, benzil ketals, and acetophenone derivatives belong to this class. Here, the photoinitiator undergoes fragmentation when exposed to UV light. The benzoyl radical is the major initiating species in the cleavage of benzoin alkyl ether. Cleavage of benzoin methyl ether is shown in Equation (19).

Benzoin alkyl ether   Benzoyl radical   Methoxy benzyl radical

$$(19)$$

(2) Radical generation by electron transfer: This mechanism involves photolytic excitation of the photoinitiator followed by electron transfer to a hydrogen atom donor, a tertiary amine [Equation (20)].

$$Ar_2C=O + \underset{R'}{\overset{R}{>}}N-CH_2-R'' \xrightarrow{h\nu} Ar_2-\overset{\cdot}{C}OH + \underset{R'}{\overset{R}{>}}N-\overset{\cdot}{C}H-R''$$

$$\underset{\substack{\text{decays to} \\ \text{inert species}}}{} \qquad \underset{\substack{\text{initiating free} \\ \text{radical}}}{}$$

(20)

Typical photoinitiators in this category are 2,2-diethoxyacetophenone, 2,2-dimethoxy-2-phenyl acetophenone, etc.

(3) Cationic type: Aryl diazonium salts $PhN_2^+X^-$ undergo fast fragmentation under UV radiation with the formation of free Lewis acids, which are known for cationic cure of epoxides [Equation (21)]

$$PhN_2^+BF_4^- \xrightarrow{h\nu} PhF + N_2 + BF_3 \qquad (21)$$

1.6.2.2 Polymer Systems

One of the early UV curable systems was based on unsaturated polyester and styrene. The unsaturation located in the polymer chain undergoes direct addition copolymerization with the vinyl group of the monomer leading to a cross-linked network (Figure 1.14).

Multifunctional acrylates are the most widely used systems. The oligomer is usually a urethane or epoxy chain end capped on both sides by acrylate groups. The molecular weight ranges from 500–3000. Reactive diluents are added to lower the viscosity of the oligomers and to increase the cure rate. The reactive diluents are generally mono- or multifunctional acrylate compounds with a molecular weight less than 500. A reaction sequence is given below (Figure 1.15).

Cationic polymerization of epoxides is another method used. As discussed earlier, Lewis acids promote ring opening polymerization of epoxides, lactones, or acetals.

The free-radical-induced polymerization is inhibited by oxygen. Several methods have been developed to reduce the undesirable effects of oxygen.

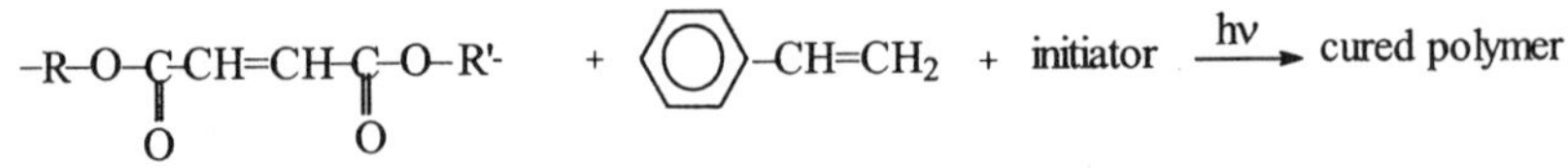

$$-R-O-\underset{\substack{\| \\ O}}{C}-CH=CH-\underset{\substack{\| \\ O}}{C}-O-R'- \quad + \quad \langle\bigcirc\rangle-CH=CH_2 \quad + \quad \text{initiator} \xrightarrow{h\nu} \text{cured polymer}$$

Figure 1.14 Unsaturated polyester system.

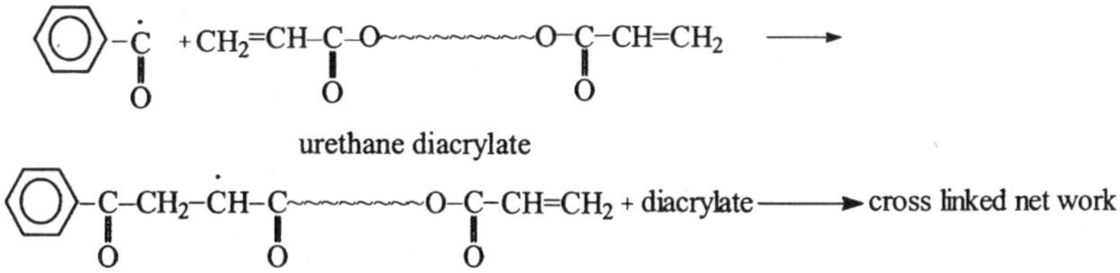

$$\text{(C}_6\text{H}_5)-\overset{\cdot}{\underset{\underset{O}{\|}}{C}} + CH_2=CH-\underset{\underset{O}{\|}}{C}-O\sim\sim\sim O-\underset{\underset{O}{\|}}{C}-CH=CH_2 \longrightarrow$$

urethane diacrylate

$$\text{(C}_6\text{H}_5)-\underset{\underset{O}{\|}}{C}-CH_2-\overset{\cdot}{CH}-\underset{\underset{O}{\|}}{C}\sim\sim\sim O-\underset{\underset{O}{\|}}{C}-CH=CH_2 + \text{diacrylate} \longrightarrow \text{cross linked net work}$$

Figure 1.15 Cross linked acrylate system.

Some of these are as follows:

- increase of UV lamp intensity
- optimization of the photoinitiator system
- curing in an inert atmosphere
- addition of oxygen scavengers

1.6.3 EQUIPMENT

(1) Ultraviolet light: different types of UV curing technologies are known. They are medium pressure mercury vapor lamp, electroless vapor lamps, and pulsed xenon lamps.

(2) Electron beam: the source of electrons is a tungsten filament that is inside a vacuum tube. The electrons are accelerated by the application of high voltage, 150,000 to 300,000 V. The accelerated electrons pass through a metallic foil window and are directed on the polymer meant for curing. X-rays are generated along with the electrons; therefore, it is necessary to shield the entire housing of the EB equipment. The energy received by this formulation is known as a dose and is termed megarad (1 Mrad = 10 joules).

1.6.4 APPLICATIONS IN TEXTILE COATING

Radiation cure has been in use in graphic arts, inks, printing, laminating, packaging, and in the electronic industry [36–38]. Motivated by the significant advantages of radiation-cured coatings over the conventional solvent-based thermal cure systems, Walsh and coworkers carried out an extensive study of radiation-cured coatings for textiles for different end uses. The work has been reported in a series of publications [39–43]. One application investigated was the backcoating of upholstery fabric [39], usually done by a thin latex coating, to stabilize the fabric against distortion and yarn raveling. A comparative study was done on nylon-viscose upholstery fabric. The latex coating was carried out by spraying and thermal curing, while the UV coating formulation was transfer coated on the fabric and cured by UV radiation. The flexural rigidity and yarn raveling tests showed that UV-cured samples were comparable to the latex-coated samples. A cost analysis showed that UV curing was only economical at low weight add-on (~2%). Studies were also conducted to achieve thicker coatings for synthetic leather applications, by radiation cure process, in lieu of the conventional solution-based polyurethane coating [40]. A suitable EB cure formulation consisting of acrylated urethane oligomer and acrylate monomer was coated on woven and knitted textile substrates by transfer coating and cured by electron beam (5 Mrads). The transfer coating was done by different ways.

None of the coated specimens, however, passed the flex fatigue test required for apparel fabric. This is largely due to the high degree of cross-linking in the cured film. A novel application studied by the authors was the simultaneous coating on both sides of nonwoven fabric by EB cure, in order to improve the durability and aesthetics of the coated fabric [41]. Simultaneous coating on both sides of the fabric is not possible in conventional solution-based processes. Essentially, the coating formulation was cast on two release papers and transfer coated on nonwoven substrate on both sides. This sandwich construction was then cured by electron beam.

Walsh et al., [42,43] continued their studies on the mechanical properties of EB-cured films using polyester acrylate urethane oligomers of different molecular weights and different reactive monomers with a view to developing suitable coatings for textiles. It was found, that higher molecular weight of the oligomer lowers the modulus of the film, the T_g, and the breaking strength. Addition of a chain transfer agent improved toughness as well as extensibility. Studies have also been carried out at TNO laboratory Holland [37,44]. They have carried out studies on UV curable coatings and the development of UV curable binders for pigment printing. Pigments have to be carefully selected so that photoinitiators and the pigment have different absorption characteristics, otherwise, insufficient curing occurs and bonding is poor.

The use of radiation cure for textiles has not yet become popular, because few systems can meet the requirements of textile coating, i.e., flexibility, strength of the film, bonding, and chemical resistance [37].

1.7 REFERENCES

1. *Rubber Technology and Manufacture,* C. M. Blow, Ed., Butterworths, 1971.
2. *Science and Technology of Rubber,* F. R. Eirich, Academic Press, New York, 1978.
3. *Rubber Technology Handbook,* W. Hoffmann, Hanser Publishers, Munich, 1989.
4. *Rubber Chemistry,* J. A. Brydson, Applied Science, London, 1978.
5. Advances in silicone rubber technology, K. E. Polmanteer, in *Handbook of Elastomers,* A. K. Bhowmik and H. L. Stephens, Eds., Marcel Dekker, 1988, pp. 551–615.

6. J. Schwark and J. Muller, *Journal of Coated Fabrics,* vol. 26, July, 1996, pp. 65–77.

7. *Encyclopedia of PVC,* vol. 1–3, L. I. Nass, Ed., Marcel Dekker Inc, New York, 1977.

8. *PVC Plastics,* W. V. Titow, Elsevier Applied Science, London and New York, 1990.

9. *Polyvinyl Chloride,* H. A. Sarvetnick, Van Nostrand Reinhold, New York, 1969.

10. *PVC Technology,* A. S. Athalye and P. Trivedi, Multi Tech Publishing Co., Bombay, 1994.

11. *Manufacture and Processing of PVC,* R. H. Burges, Ed., Applied Science Publishers, U.K, 1982.

12. *Polyurethane Handbook,* G. Oertel, Hanser Publishers, Munich, 1983.

13. *Polyurethane Elastomers,* C. Hepburn, Applied Science Publishers, London and New York, 1982.

14. *Polyurethane Chemistry and Technology,* Pt. I and II, J. H. Saunders and K. C. Frisch, Interscience Publisher, 1964.

15. *Developments in Polyurethanes,* J. M. Buist, Applied Science Publishers, London, 1978.

16. *Thermoplastic Polyurethane Elastomers,* C. S. Schollenberger, in *Handbook of Elastomers,* A. K. Bhowmik and H. L. Stephens, Eds., Marcel Dekker, New York, 1988, pp. 375–407.

17. R. Heath, *Journal of Coated Fabrics,* vol. 15, Oct., 1985, pp. 78–88.

18. J. W. Rosthauser and K. Nachtkamp, *Journal of Coated Fabrics,* vol. 16, July, 1986, pp. 39–79.

19. J. T. Zermani, *Journal of Coated Fabrics,* vol. 14, April, 1985, pp. 260–271.

20. J. T. Tsirovasiles and A. S. Tyskwicz, *Journal of Coated Fabrics,* vol. 16, Oct., 1986, pp. 114–121.

21. J. Goldsmith, *Journal of Coated Fabrics,* vol. 18, July, 1988, pp. 12–25.

22. F. B. Walter, *Journal of Coated Fabrics,* vol. 7, April, 1978, pp. 293–307.

23. H. G. Schmelzer, *Journal of Coated Fabrics,* vol. 17, Jan., 1988, pp. 167–181.

24. R. W. Oertel and R. P. Brentin, *Journal of Coated Fabrics,* vol. 22, Oct., 1992, pp. 150–159.

25. J. R. Damewood, *Journal of Coated Fabrics,* vol. 10, Oct., 1980, pp. 136–150.

26. Polyurethane structural adhesives, B. H. Edwards, in *Structural Adhesives Chemistry and Technology,* S. R. Hartshorn, Ed., Plenum, New York, 1986, pp. 181–215.

27. C. S. Schollenberger and F. D. Stewart, *Advances in Urethane Science and Technology,* vol. 2, 1973, pp. 71–108.

28. Surface Analysis and Pretreatment of Plastics and Metals, D. M Brewis, Macmillan, New York, 1982.

29. N. K. Porter, *Journal of Coated Fabrics,* vol. 21, April, 1992, pp. 230–239.

30. *Textile reinforcement of Elastomer,* W. C. Wake and D. B. Wootton, Applied Science Publishers Ltd., London, 1982 (and references therein).

31. A. G. Buswell and T. J. Meyrick, *Rubber Industry,* Aug., 1975, pp. 146–151.

32. B. T. Knight, *Journal of Coated Fabrics,* vol. 21, April, 1992, pp. 260–267.

33. Coated fabrics, B. Dutta, in *Rubber Products Manufacturing Technology,* A. Bhowmik, M. M. Hall and H. A. Benary, Eds., Marcel Dekker, New York, 1994, pp. 473–501.

34. Vulcanization and curing techniques, A. K. Bhowmik and D. Mangaraj, in *Rubber Products Manufacturing Technology,* A. K Bhowmik, M. M. Hall and H. A. Benary, Eds., Marcel Dekker, New York, 1994, pp. 315–394.

35. Radiation cured coatings, V. Koleske, in *Coating Technology Handbook,* D. Satas, Ed., Marcel Dekker, New York, 1991, p. 659.

36. C. Decker, *Journal of Coating Technology,* vol. 59, no. 751, Aug., 1987, pp. 97–106.

37. E. Krijnen, M. Marsman and R. Holweg, *Journal of Coated Fabrics,* vol. 24, Oct., 1994, pp. 152–161.

38. C. Bluestein, *Journal of Coated Fabrics,* vol. 25, Oct., 1995, pp. 128–136.

39. W. K. Walsh, K. Hemchandra and B. S. Gupta, *Journal of Coated Fabrics,* vol. 8, July, 1978, pp. 30–35.

40. B. S. Gupta, K. Hemchandra and W. K. Walsh, *Journal of Coated Fabrics,* vol. 8, Oct., 1978, pp. 183–196.

41. B. S. Gupta, W. S. McPeters and W. K. Walsh, *Journal of Coated Fabrics,* vol. 9, July, 1979, pp. 12–24.

42. W. Oraby and W. K. Walsh, *Journal of Applied Polymer Science,* vol. 23, 1979, pp. 3227–3242.

43. W. Oraby and W. K. Walsh, *Journal of Applied Polymer Science,* vol. 23, 1979, pp. 3243–3254.

44. A. H. Luiken, M. P. W. Marsman and R. B. M. Holweg, *Journal of Coated Fabrics,* vol. 21, 1992, pp. 268–300.

CHAPTER 2

Textile Substrate for Coated Fabric[1]

2.1 MATERIALS AND TRENDS

A wide range of textile materials is used as substrates for coated fabrics. These may be woven, knitted, or nonwoven materials. The importance of textile materials can be gauged from the use of several billion square meters of fabric every year.

The types of fiber commonly used in coating are cotton, rayon, nylon, polyester, and blends of polyester with cotton or rayon, depending on the end use requirements. Polyester is the most popular in staple form for nonwoven material and in spun form for woven material. Polypropylene is emerging as the fiber of choice because of its low specific gravity, strength properties, chemically inert nature, and low cost. However, its poor dyeability, adhesion, and thermal stability are disadvantages that need to be overcome. High performance fibers like Kevlar®, Nomex®, PBI, etc., are used in specialized applications.

In woven form, plain, basket, twill, and sateen constructions are generally used. Among the knitted constructions, circular knits are used as a substrate for upholstery fabric. Warp knit fabrics, particularly weft inserted warp knits (WIWK), are preferred for making coated fabrics for special applications. Non-woven fabrics, produced by different techniques, find use in sanitary and medical products, apparel, artificial leather, dot-coated fabrics for fusible interlinings, etc.

The emerging trends in the use of textiles can be summarized as follows [1]:

(1) Development of polyester fiber with lower elongation or higher modulus, higher adhesion, and microdenier filament for greater cloth cover/surface area

(2) Greater use of polypropylene

---

[1] This chapter was contributed by N. Kasturia, R. Indushekhar, and M. S. Subhalakshmi, DMSRDE, Kanpur, India.

(3) Use of longer roll length and wider fabric to lower the cost

(4) More use of textured, Dref, and core spun yarns for improved adhesion

(5) Greater use of nonwoven and WIWK

The choice of proper fabric for coating is as important as the selection of the polymer, because it offers the primary physical property to the end product. For proper selection of fabric, the following aspects need to be considered:

- strength and modulus
- creep behavior
- resistance to acids and chemicals
- adhesion requirement
- resistance to microbiological attack
- environment of use
- durability
- dimensional stability
- cost

The following characteristics should be considered when designing a textile substrate to meet specific end use requirements:

(1) Fiber type and form such as staple, filament, etc.

(2) Yarn type and construction

(3) Fabric form, i.e., woven, nonwoven, and knitted and their construction

2.2 TEXTILE FIBERS

The textile fabric/substrate used for coating is made of textile fibers. There are two main types of fibers: natural fibers and man-made or synthetic fibers. The natural fibers may be of vegetable origin, such as cotton, kapok, flax, coir, sisal, etc.; of animal origin such as wool, silk, etc.; and of mineral origin, such as asbestos. The vegetable fibers are cellulosic in nature, the animal fibers are proteins, and asbestos is a silicate. The organic man-made fibers are essentially of two types: derived from cellulose, such as rayon and acetate, and synthetic polymers, such as nylon, polyester, acrylics, polypropylene, etc. Metallic fibers and glass fibers are inorganic man-made fibers. The properties of some important fibers used in the coating industry are discussed in this section. The physical and chemical properties of the fibers have been summarized in Tables 2.1 and 2.2, respectively.

2.2.1 COTTON

It is known as the king of fibers. Cotton is a cellulosic ($\sim$94% cellulose) staple fiber. The fiber length varies from 10–65 mm, and fiber diameter ranges

TABLE 2.1. Important Physical Properties of Fibers.

Fibers / Properties	Cotton	Rayon Viscose	Rayon Acetate	Nylon Normal apparel grade	Nylon Industrial grade	Polyester Normal Staple Fiber	Polyester High tenacity fiber	Polypropylene Multifilament fiber	Polypropylene Staple fiber	Polypropylene High tenacity fiber	Aramid (Nomex®)
Specific gravity	1.52–1.55	1.52	1.32	1.14	1.14	1.36	1.36	0.90	0.90	0.90	1.38
Tensile strength g/d	3–5	2.6	1.4	4.1–5.5	6.3–8.18	3.5	9.5	5–7	4–6	5.5–8.5	5.3
Elongation at break %	4–13	10–30	25–50	26–32	14–22	10–40	—	15–35	20–35	15–25	22
Moisture regain % at 21°C, 65% RH	8.5	≅13	6.3–6.5	4	4	0.4	0.4	negligible	negligible	negligible	5–5.2
Effect of heat (a) Resistant temperature	150°C	150°C	150°C	180°C	180°C	180°C					370°C
(b) Decomposition temperature	230°C	210°C	210°C			—					~500°C
(c) Melting temperature	Decomposes	Decomposes	Decomposes	250°C (Nylon 66) 215°C (Nylon 6)	—	250°C	250°C		160–175°C		Decomposes

TABLE 2.2. Important Chemical Properties of Fibers.

Fibers Properties	Cotton	Rayon	Nylon	Polyester	Polypropylene	Aramid (Nomex®)
Effect of sunlight and atmosphere	Loss of tensile strength and discoloration of fibers occur	Loss of tensile strength	Appreciable degradation by sunlight	Low degradation in shade. Direct sunlight weakens fibers	Rapid degradation to sunlight and weathering.	Resistance to aging is excellent
Effect of microorganism	Mildew, microorganisms degrade the fiber	More resistant than cotton	Resistant	Resistant	Resistant	Resistant
Effect of acids	Deteriorates the fiber. Mineral acids degrade more readily than organic acids	Same as cotton	Affected by concentrated mineral and organic acids	Resistant to most mineral acids. Concentrated sulfuric acid decomposes fiber	Excellent resistance to acids	Not significantly affected, but is attacked by boiling sulfuric acid
Effect of alkalis	Resistant at room temperature but swelling occurs	Same as cotton	Virtually no effect	Resistant to alkali at room temperature but hydrolytic degradation occurs at boiling temperature	Resistant to alkalis	Resistant to alkalis
Effect of solvents/ oxidizing agents	Resistant to common hydrocarbon solvents. Oxidizing agents convert it to oxycellulose	Same as cotton	Benzene, chloroform, acetone, and ether do not affect, but it dissolves in phenols and strong acids	Resistant to hydrocarbon solvents. Soluble in m-cresol, o-chlorophenol at high temperature	Insoluble in organic solvents at room temperature. Dissolves in hot decalin, tetralin. Attacked by oxidizing agents	Resistant to most organic solvents

from 11–22 μm, respectively. The fiber has good strength due to the large number of interchain hydrogen bonds present in the polymer chain. Cotton is a natural fiber with wide variation in properties. This variation is caused by differences in climatic conditions in the regions where the cotton is grown. A good quality cotton fiber is characterized by its finer fiber diameter and longer staple length. The important commercial varieties are (1) Sea Island, (2) Egyptian, (3) American upland, and (4) Indian cotton. Sea Island and Egyptian cotton have higher staple length and produce finer quality yarns. Indian cotton has shorter fiber length and produces coarser yarns. American upland cotton lies between these two categories for quality and fiber length.

Cotton has moderate mechanical strength when dry but good wet strength. The resiliency of the fiber is low, therefore, cotton fabrics wrinkle easily. Due to high moisture absorption of the fiber, cotton fabrics are comfortable as summer wear. Cotton is extensively used for apparel fabrics as well as industrial textiles like canvas, ducks, etc. The fabric has excellent adhesion to coated/laminated polymeric film.

2.2.2 RAYON

Rayons are man-made fibers derived from cellulose. Viscose rayon is regenerated cellulose, while acetate rayon is obtained by acetylation of cellulose. Both the fibers are characterized by high luster and are considered as artificial silk. Viscose is obtained by treating wood pulp with caustic soda solution to form soda cellulose. It is then treated with carbon disulfide to form cellulose xanthate solution. The alkaline cellulose-xanthate, is ripened, and on achieving the required viscosity, the solution is spun into a coagulating bath of dilute (10%) sulfuric acid. The viscose filament gets precipitated there. For making acetate rayon, wood pulp or cotton linters are treated with a solution of acetic anhydride in glacial acetic acid to form secondary cellulose acetate, which has fiber-forming properties. The secondary cellulose acetate is made into a dope with acetone. The dope is forced through holes of a spinnerette, and the filament is solidified by evaporation of acetone in hot air.

Like cotton, rayons are cellulosic in nature, as such, their chemical and physical properties are similar to those of cotton. It is used in blends with polyester for apparel fabrics, household textiles like furnishings and carpets, and in medical fabrics.

2.2.3 NYLONS

Nylon is the common name of linear aliphatic polyamides. The most important fibers in this class are nylon 66 and nylon 6. Nylon 66 is polyhexamethylene adipamide, a condensation polymer of hexamethylene diamine and adipic acid. The suffix 66 stands for the number of carbon atoms in the monomers.

Nylon 6 is polycaprolactamide, the monomer being ε-caprolactam. The reaction sequences are given below.

$$n\text{H}_2\text{N(CH}_2)_6\text{NH}_2 + n\text{HOOC(CH}_2)_4\text{COOH} \longrightarrow \text{-[NH(CH}_2)_6\text{-NH-CO-(CH}_2)_4\text{-CO]}n\text{-}$$

Hexamethylene diamine   Adipic acid   Nylon 66

$$\begin{bmatrix} -(\text{CH}_2)_5- \\ -\text{NH-}\ \text{CO-} \end{bmatrix} \longrightarrow \text{- [NH(CH}_2)_5\text{-CO-]}n\text{-}$$

ε-Caprolactam   Nylon 6

Nylon is a group of synthetic super polymers, with much higher strength and elongation than cellulosic fibers. It is available as regular translucent fine filament and can be converted into staple fibers.

Being a thermoplastic material, nylon fabric undergoes thermal shrinkage, and besides, it generates static electricity on friction. Special precautions therefore have to be taken while processing nylon fibers. Nylon fabrics are widely used for carpets, upholstery, and apparel. The high strength, elasticity, and abrasion resistance enables nylon to be used for a variety of industrial end uses such as filter fabrics, nets, webbings, cordages, parachutes, ropes, ballistic fabrics, etc.

2.2.4 POLYESTER

Polyester refers to a class of polymers containing a number of repeat ester groups in the polymeric chain. Commercially available polyester fiber is polyethylene terephthalate. It is known in different countries by different brand names. In the U.K., it is known as Terylene, and in the U.S., it is known as Dacron. The fiber is available in filament as well as in staple fiber form. A number of other polyesters have been converted into fibers, but they have not been exploited commercially.

$$n\text{CH}_3\text{OOC-}\langle\bigcirc\rangle\text{-COOCH}_3 \ + \ n\text{HO-CH}_2\text{-CH OH} \longrightarrow$$

Dimethyl terephthalate   Ethylene glycol

$$\text{H}\begin{bmatrix} -\text{OCH}_2\text{CH}_2\text{OOC} -\langle\bigcirc\rangle\text{-CO-} \end{bmatrix}_n\text{-OCH}_2\text{CH}_2\text{OH}$$

Polyethylene terephthalate

Like nylon, polyester fabrics also generate static electricity and undergo thermal shrinkage. Fabrics show poor adhesion to the coated polymeric film.

The major use of polyester and its blends with cotton, rayon, and wool are in apparel fabrics, household fabrics, and industrial textiles.

2.2.5 POLYPROPYLENE FIBER

Polypropylene is a hydrocarbon fiber, the properties are dependent on the microstructure of the fiber. From the textile point of view, only isotactic polypropylene can be fibrillated, and the isotacticity index should be higher than 90%. The average molecular weight of polypropylene fiber ranges from 100,000–300,000. Polypropylene fibers are produced in different forms like staple, monofilament, and multifilament.

Due to its light weight, negligible water absorption, and high abrasion resistance, polypropylene is widely used for making ropes, fishing nets, tufted carpets, etc.

2.2.6 ARAMIDS

These are aromatic polyamides that are closely related to nylons. In aramids, the aliphatic carbon chain is replaced by aromatic groups, bringing about considerable change in the properties of the resultant fiber. The first fiber introduced in this class by Dupont U.S.A. was Nomex®, which is chemically poly-*m*-phenylene isophthalamide, a condensation product of m-phenylene diamine and isophthalic acid. Nomex® is flame resistant and is widely used for fireproof clothing. The *p* isomer, viz., polyparaphenylene terephthalamide (PPT) is known as Kevlar® fiber which possesses ultrahigh strength and modulus. The properties of Nomex® are given in Tables 2.1 and 2.2.

Aramids are high strength and high modulus fibers. They are mainly used in composite reinforcement for ballistic protection, ropes, cables, and for fire-resistant clothing.

2.3 SPINNING

Spinning refers to the process of conversion of small fibers into yarns, or in case of synthetic fibers, spinning refers to the processes that convert polymers into filaments. Most of the natural fibers like cotton, wool, etc., are available only as staple fibers having different fiber lengths. Spinning of natural fibers is divided into the following systems depending upon the fiber lengths:

a. Short staple spinning system or cotton spinning system

b. Long staple spinning system or wool spinning system

Synthetic filaments, when converted to staple fibers, are spun by a process similar to that of cotton or wool.

2.3.1 COTTON SPINNING SYSTEM

Cotton, which is a vegetable fiber, undergoes a sequence of processes before being spun into yarn. This sequence includes ginning, opening and cleaning (blow room), carding, lap formation (sliver and ribbon lap), combing (optional), drawing, roving, and, finally, spinning. The flowchart (Figure 2.1) gives an overview of the sequential operations involved in cotton spinning.

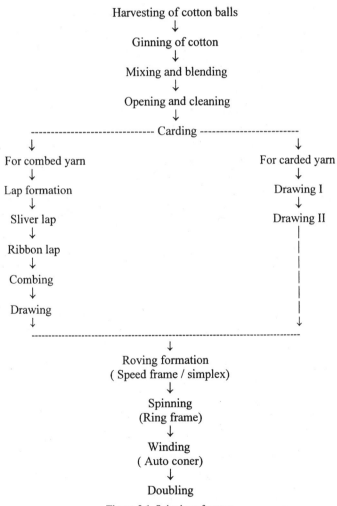

Figure 2.1 Spinning of cotton.

2.3.1.1 Ginning

Ginning is the starting process to which cotton is subjected on its way from the field to the textile mill before it is spun into yarn. Ginning is the process of separating cottonseed from the fiber. During this process, foreign matter like leaf bits, stalks, hulls, etc., are removed. Care is taken to preserve the quality of fiber, particularly the fiber length. The cotton fibers removed from the seed are compressed into large bales and sent to mills for processing.

2.3.1.2 Blow Room

In a textile mill, this is the first preparatory process. In the blow room, cotton bales are subjected to mixing, opening, and cleaning processes. The cotton bales from different farm fields are fed into a mixing bale opener (MBO). Mixing of different varieties of cotton is done to improve the uniformity, thereby improving the quality and minimizing the raw material cost. In some cases, cotton fibers are blended with other fibers to manufacture special yarns with desired properties. The MBO opens the compressed cotton and mixes the different varieties of cotton by rotating cylindrical beaters. Thus, the closely packed fibers are loosened, and during this process, dirt and other heavy impurities are separated from the fiber, either by gravity or by centrifugal force. The loosened fibers are converted into a lap of smaller tufts called flocks.

2.3.1.3 Carding

The fibers are received at the carding machines either in lap form or as flocks. Lap and flock feeding have their own advantages and disadvantages. However, most modern mills have flock feeding systems. The main objectives of the carding process are to continue the cleaning process, removing some amount of short fibers, fiber individualization, partially aligning the fibers in the direction of the fiber axis, and disentangling neps (small entangled collection of immature fibers). In the carding zone, the fibers pass over the main cylinder. The main cylinder is made of cast iron and is 120–130 cm diameter. This cylinder is covered with fine sawtooth wires. On the top of the cylinder, there are a number of moving flats that are joined to form an endless, circulating band. The flats are cast iron bars with one side clamped with a clothing strip, which is rubberized fabric fixed with angled steel wires. These flats and main cylinder together form the main carding zone. The carding action involves the transfer of fibers from the cylinder surface to the flat surface and vice versa. During the multiple transfer, the wire points in the cylinder try to retain the fiber, and at the same time, the wire points in flats try to pluck the fiber. But, most of the fibers are retained in the cylinder wire points, as the flats rotate at a much slower speed than the main cylinder. This process leads to fiber individualization.

2.3.1.4 Lap Formation

The carding process opens up the collected mass of fibers so that the fibers become individual. However, the fibers in the card sliver are not completely aligned or oriented in the fiber axis. Some fibers lie haphazardly in the sliver. Thus, the card sliver is given a minimum of two drafting processes before it goes to the next machine. In this process, the sliver is passed between sets of rollers that are running at different speeds, each succeeding pair rotating faster than the previous so that the fibers are pulled in a lengthwise direction. These two drafting operations are achieved by the sliver lap and ribbon lap machines. To improve the uniformity of the sliver, it is subjected to the process called doubling: Doubling is the process of combining a number of slivers. By this process, the thin and thick places present in the sliver are evened out. In the sliver lap machine, 16–20 card slivers are creeled and passed through the feed table to three pairs of drafting rollers for the drafting operation. The drafted slivers are then taken to two pairs of calender rollers that compress the sliver material. This drafted and compressed sliver material called lap is wound finally on a spool.

2.3.1.5 Combing

Combing is an optional process, which is introduced into the spinning of finer and high quality yarns from finer cotton. For coarser cotton fibers, the combing operation is usually omitted. This is the process of removal of a predetermined length of short fibers present in the fiber assembly, because the presence of short fibers reduces the yarn quality by increasing the number of thin and thick places, neps, and hairiness, and also lowers the tenacity. The presence of short fibers and the inappropriate configuration of the fibers in the drawn sliver would not allow drafting and the ring frame operations to be effective. Thus, combing is an important process next to carding for spinning fine yarns.

In the combing operation, lap from the lap roller is unwound and fed to the nippers by the feed rollers periodically. The fiber material is gripped between the top and bottom nippers that keep the material ready for rotary combing. The rotary comber is a cylindrical device having needles fitted in a part of its surface. The comber needles enter the fringe and comb and straighten the fibers. During this operation, short fibers, which are not under the grip of the nippers, are combed away with the needles. Drafted slivers are finally delivered into the can. In a comber, there are eight feeding heads. In each head, one lap is fed, and the comber output is in the form of a sliver.

2.3.1.6 Drawing

Sliver is taken from combing machine to the drawing machine. The main objectives of the drawing process are to further straighten the fibers, make them parallel to the fiber or sliver axis, and improve the uniformity by doubling. Blending of two different fibers like polyester and cotton, polyester and viscose, etc., is also carried out in draw frame in the case of manufacturing blended yarn. Fiber straightening is achieved by drafting the slivers. During the drafting process, the linear density (weight per unit length) of sliver is also reduced. The operating principle of draw frame is that four to eight card/comber slivers are fed to the drafting arrangement through feed rollers, which are carried in a creel frame or table. The drafted slivers come out as an even web that is immediately condensed into a sliver to avoid disintegration of the web by a converging tube.

2.3.1.7 Speed Frame/Fly Frame

The uniform sliver obtained from the draw frame, subsequently goes to the speed frame, which is the final machine in the spinning preparatory operations. The main tasks here are attenuation of fibers and formation of a suitable intermediate package. Attenuation is the reduction in the linear density of the sliver. The extent of reduction is such that it is suitable for spinning into a yarn. The attenuation of the sliver is achieved by drafting. By this drafting operation, the sliver becomes finer and finer, and the resultant product is called the "roving." After the drafting operation, the roving is wound on the bobbin. During winding, a little amount of twist is imparted to the roving.

2.3.1.8 Ring Spinning

The final process of yarn formation, i.e., spinning, is carried out in the machine called a ring frame. In this process, the roving is attenuated into yarn by drafting. Substantial amount of twist is inserted to the yarn, then it is wound on a bobbin. In other words, drafting, twisting, and winding are the steps taking place during the spinning operation. In the spinning process, the roving, which is several times thicker than the yarn, is subjected to a higher amount of draft when it passes through three pairs of closely associated rollers moving at different surface speeds. For attenuation, the yarn delivery rollers revolve at a higher speed than the feed rollers.

The yarn produced from the spinning machine is a single yarn. However, as per the end use requirement, it may be twisted together with two or more single yarns in the doubling machine to achieve stronger and more uniform yarn.

2.3.1.9 Newer Methods of Spinning

Over the centuries, many ways have been devised for conversion of fibers into yarns, but in the past thirty years, the search for a new and more economical spinning system has been actively pursued in many parts of the world. A new spinning system popularly known as rotor spinning or open-end spinning was introduced in the late 1960s. It has made great impact on the textile industry, especially in terms of rate of production. Other spinning processes, like electrostatic spinning, air-vortex spinning, friction spinning, and disc spinning, are also different types of open-end spinning processes. The twist spinning, self-twist spinning, wrap spinning, false twist spinning, and adhesive processes are only of academic interest and have not become popular. All of these new spinning systems produce yarns having quality that differs to a certain extent from that produced by the more traditional ring-spinning process.

2.3.2 SYNTHETIC FIBER SPINNING

Synthetic fiber spinning is entirely different from staple fiber spinning which was discussed earlier. In synthetic fiber spinning, the fiber/filament is made by extruding the polymer liquid through fine holes. In synthetic fiber spinning, the diameter of the filament is determined by three factors, i.e., the rate at which the dope is pumped through the spinnerette, the diameter of the spinnerette holes, and the rate at which they are taken up by the take-up rollers.

Synthetic fiber spinning is divided into three systems based on the meltability and solubility of the polymer. They are melt spinning and solution spinning. Solution spinning may be further divided into two systems on the basis of nature of the solvent: dry spinning and wet spinning.

2.3.2.1 Melt Spinning

Polymers that melt on heating without undergoing any decomposition are spun by a melt spinning system. In this system of spinning, the polymer chips are fed into a hopper. From the hopper, the chips are passed to a spinning vessel through a pipe. In the spinning vessel, the polymer chips fall onto an electrically heated grid that melts the chips and has a mesh too small to pass the chips until they are melted. The molten polymer then passes into the pool and filtering unit. The filtering unit consists of several layers of metal gauge and sand kept alternately with coarse, fine, and very fine mesh and particle size, respectively. This filtering unit filters any impurities out of the molten polymer mass, as they may block the fine holes present in the spinnerette plate. After passing out of the filtering unit, the polymer is forced through the holes in the spinnerette plate and emerges from the plate. As the filaments emerge, they are drawn away from the outlet, stretching the polymer before it cools. Immediately after emerging, cool air is passed to solidify the melt. The solidified melt, now called filament, is then passed through a spin finish bath, an antistatic agent is added, and the filament is wound onto a bobbin by a winder. The filament is then used as such or is imparted crimped or cut into staple fibers. Polyester, nylon, and polypropylene polymers are converted into filaments by this technique.

2.3.2.2 Solution Spinning

Some polymers, e.g., acrylics, undergo decomposition on heating. Therefore, they cannot be spun by the melt spinning system. These polymers are spun by a solution spinning system. In this system, the polymer is dissolved in a suitable solvent, and the polymer solution, generally called dope, is extruded through the holes of the spinnerette for making filaments. If the solvent selected is a volatile solvent, then the polymer is spun by a dry spinning system; if it is a nonvolatile solvent, the polymer is spun by a wet spinning system. Polyacrylo nitrile (PAN) polymer (acrylic fiber) can be spun by either a dry or wet spinning system.

In a dry spinning system, the polymer is dissolved in a suitable volatile solvent and forms a solution called dope. This dope is fed to the spinning head from a feed tank through pipes. A metering pump controls the constant and uniform flow through the spinnerette. The extruded stream of solution flows out into a hot air chamber. On evaporation of the solvent by hot air, the solidified polymer filament is drawn, taken up through a spin finish bath, and wound. The solvent may be recovered from air by adsorption on active carbon.

In wet spinning, the polymer is dissolved in a nonvolatile solvent, and the polymer solution is regularly fed to the filter and spinning head. The spinnerette is submerged in a coagulation bath, and as the polymer emerges out of the spinnerette, the polymer in the solution is precipitated by the bath liquid and soldifies in filaments. The filament is then wound onto the bobbin after spin finish application.

2.4 WOVEN FABRICS

The process of converting a set of yarns into a fabric, on a loom, is called weaving. The mechanism of interlacing two sets of yarns at right angles to each other, according to a desired design, is done on the loom. Woven fabrics are more widely used in apparel and industrial applications. The two sets of yarns, warp (longitudinal thread), and weft (lateral thread) require a separate set of processing before they are ready to be woven on the loom. This becomes pertinent, especially if one is looking for special properties like rib effect, absorbancy, and adhesion properties of the fabric. The properties of a gray fabric (fabric coming out from a loom) depend on fiber properties, yarn properties, density of yarns in the fabric, weave, and yarn crimp.

2.4.1 WEAVING

To produce a fabric on any loom, the five operations given below are necessary. The first three operations are generally termed fundamental operations.

(1) Shedding: the separation of warp threads (longitudinal) into two layers, one set of which is lifted and the other which is lowered to form a space sufficient enough to send a shuttle of weft yarn (lateral yarn) for interlacement. Each specific set of warp yarns is raised by means of a harness or heald frame. The design of the weave depends on the sequence of raising of the set of yarns forming the shed during insertion of the filling yarn. Tappets, cams, dobby, and Jacquard mechanisms are used as shedding devices.

Tappet and cams can handle up to fourteen different harnesses and are widely used for simple fabrics. Dobby is a shedding device placed on top of the loom that can handle up to forty harnesses and is used for producing small, figured patterns. Jacquard device is also placed on the top of the loom and can handle individual warp yarns. This enables the weaving of complicated and elaborate designs.

(2) Picking: the insertion of weft yarn by passing from one end of the fabric to the other end through the shed created due to parting of warp yarns into upper and lower layers. Generally, shuttles, projectiles, rapiers, etc., are used as vehicles for the transfer of weft.

(3) Beating up: pushing the newly inserted weft (pick) into the already woven fabric to the end point (fell) is known as the beating process. The beating force employed has a significant influence on the closeness of the fabric.

(4) Warp let off: delivering the series of warp threads simultaneously at a required rate at a suitable constant tension is termed as warp let off. The rate of releasing the warp threads in conjunction with the take up of cloth decides the pick density in the fabric.

(5) Cloth take up: moving the fabric from the formation zone at a constant rate and winding the fabric onto a roller is called cloth take up. By controlling the warp tension, the let off motion decides the crimp in the threads.

2.4.2 FUNDAMENTAL WEAVES

The fundamental weaves are plain, twill, and satin weaves. These are the basic weaves from which many new kinds of weaves are derived. The smallest unit of design that appears repeatedly in a weave pattern is called the repeat. The weaves that are generally used in the coating industry are discussed below.

2.4.2.1 Plain Weave

Plain weave is the simplest form of interlacing two sets of yarns. The yarns interlace each other at right angles in alternate order. It has the smallest number of yarns in the repeat, which is two. The maximum possible number of intersections of warp and weft yarns makes a plain weave fabric the strongest and stiffest among the various woven structures. About 40% of all fabrics produced are in plain weave. Some examples of plain weave fabrics are voile, muslin sheeting, mulmul, poplin, cambric, lawn, organdy, shantung, taffeta, canvas, etc. Apart from the plain weave, derivatives of plain weave (weave construction based on plain weaves) are widely used in various industrial fabrics, e.g., tents/shelters, protective clothing, parachutes, and other specialized clothing.

The derivatives of plain weave are as follows:

(1) Basket weave: this is a variation of the plain weave that uses two or more warp yarns simultaneously interlaced with two or more fillings, giving a balanced structure to produce a design that resembles the familiar pattern of a basket. They are woven in a pattern of 2 × 2, 3 × 3, or 4 × 4 with two or more filling yarns interlaced with a corresponding number of warp yarns.

(2) Oxford weave: it varies slightly from the regular basket weave in that it has 2 × 1 construction, i.e., one filling yarn passes alternately over and under two warp yarns that act as one thread. Generally, the fineness of the weft yarn is approximately equivalent to the fineness of the warp yarns.

The basket/mat weave consists of a fewer number of interlacings per cm compared to plain weave, and hence, it allows more threads to be inserted per cm. For this reason, the cloth cover of basket weave is high compared to basic plain weave, but due to fewer intersections/cm, this fabric is more flexible and drapes (hangs) well. In applications where tear strength is important, basket weaves are preferred to plain weave.

2.4.2.2 Twill Weaves

In twill weave, the first warp yarn interlaces with the first weft yarn, the second warp yarn with the second weft yarn, the third warp yarn with the third weft yarn, and so on up to the end of the repeat. Owing to this order of warp and weft yarns interlacing, fabrics with a twill weave pattern exhibit a diagonal stripe directed at an angle of 45° (diagonal lines) from the left upward to the right. These weaves are employed for the purpose of ornamentation and to make the cloth heavier and have better draping quality than that which can be produced with the same yarns in a plain weave. Twill lines are formed on both sides of the cloth. The direction of diagonal lines on the face side of cloth is opposite to that on the back side, coinciding, respectively, with the weft and warp floats on the other side. Thus, if the warp floats predominate on one side of the cloth, weft floats will predominate on the other side of the cloth. The twill weave fabrics include canton flannel, covert cloth, denim, drill, gabardine, jean, khaki, whipcord, etc.

2.4.2.3 Satin and Sateen Weaves

In a satin weave, the warp skips a number of weft yarns before interlacement, thus, warp yarn dominates the face of the fabric. On the other hand, in a sateen weave, the weft yarn skips a number of warps prior to interlacement, and weft dominates the fabric face. For example, when a warp skips seven fillings before it interlaces, the weave is termed an eight float satin. Because of the long yarn floats, satin and sateen fabrics reflect more light and impart high gloss to the surface. These fabrics are also characterized by a maximum degree of smoothness. Satin weave fabrics drape well because the weave is heavier than the twill weave, which in turn, is heavier than the plain weave. Some satin weave fabrics include antique satin, bridal satin, cotton satin, etc.

Each weave can be presented in a square paper design (point paper design) to illustrate weave pattern. Vertical columns of squares on a point paper represent warp ends, and horizontal rows of squares represent picks. A marked square on the point paper indicates that the warp end is raised above the pick, while a blank square means that the warp end is lowered under the pick during weaving. Figure 2.2 shows the graphic symbol of some basic weaves.

2.4.3 LOOMS

All woven cloth is made on some sort of loom. The conventional looms are shuttle looms. The shuttle that carries the yarn through the shed of warps is a wooden boat-like container about 30 cm long that carries a bobbin called pirn onto which filling yarn is wound. The shuttle is propelled through the shed formed by warps carrying the filling yarn across the width of the material as it unwinds from the bobbin. The conventional shuttle looms consume a great deal of power and are relatively slow in operation, noisy, and not satisfactory for wide fabrics. Because of these drawbacks, shuttleless looms have been developed. In these looms, the yarn is carried directly from a large cone of yarn located outside of the loom. The operations of some important shuttleless looms are discussed below.

(1) Projectile looms: the picking action is accomplished by a series of small bullet-like projectiles that grip the filling yarn and carry it through the shed and then return empty. These looms have speeds between 300–600 picks per minute (ppm) depending on the width of the fabric.

(2) Rapier looms: instead of the projectile, a rapier-like rod or steel tape is used in these looms to carry the filling yarn. More commonly, two rods are

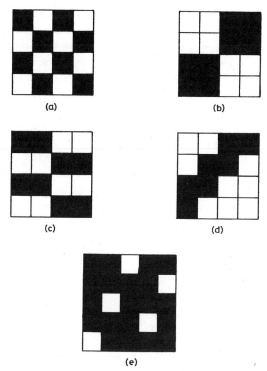

Figure 2.2 Graphic symbol of some basic weaves: (a) plain weave, (b) 2 × 2 basket weave, (c) 2 × 1 basket/oxford weave, (d) 2 × 2 twill weave, and (e) 4 × 1 satin weave.

used: one carries the yarn halfway, where the end of the yarn it carries is transferred to a rod propelled from the other side that pulls the yarn the rest of the way, while the first rod retraces. This speeds the operation.

(3) Water-jet looms: in water-jet looms, a predetermined length of pick is carried across the loom by a jet of water propelled through the shed. These looms operate at high speeds of ~600 ppm. These looms can produce superior quality fabrics.

(4) Air-jet looms: a jet of air is used to propel the filling yarn through the shed at a relatively high speed of ~600 ppm.

2.5 KNITTED FABRICS

Knitting is a process whereby fabrics are formed by the interlacing of neighboring yarn loops. The fabric manufactured by knitting has distinctly different properties than those of the woven structures. The knitted structure may be formed either (a) by weft knitting, in which one or more individual weft supply yarns are laid across beds of needles so that loops of yarns are drawn through previously made loops, or (b) by warp knitting, in which the fabric is formed by looping together parallel warp yarns as they are fed collectively from a warp beam. In a knitted fabric, the yarn density is denoted by wales and courses per cm. The wales are a series of loops in successive rows lying lengthwise in the fabric; they are formed by successive knitting cycles and the intermeshing of each new loop through the previously formed loop. It gives an indication of needles per cm in the machine. Courses are the horizontal ridges in the weft direction that give an idea of the stitch length. A stitch is made when a loop of yarn is drawn through a previously made loop. Due to fewer manufacturing steps, knitted fabrics are easy to produce compared to woven fabric. Moreover, the changeover from one structure to the other can be readily done. Knitted fabrics have the following important characteristics:

- high extensibility
- shape retention on heat setting
- crease and wrinkle resistance
- pliability
- better thermal insulation property
- better comfort property

These fabrics find wide application in casual wear, sportswear, and under-garments. In the coating industry, they are widely used for making upholstery fabric and leather cloth.

2.5.1 WEFT-KNITTED STRUCTURES

Weft knitting process is the method of creating a fabric via the interlocking of loops in a weftwise or crosswise direction. The three most popular and fundamental structures of weft-knitted structures are jersey (plain), rib, and purl.

In jersey-knit fabrics, the vertical component of the loops appears on the face side, and the horizontal component is seen on the reverse side of the fabric. The face side of jersey usually has a softer hand than the reverse side. The fabric is characterized by a smooth, regular surface with visible wales on the face side, and a series of semicircular loops on the reverse side. The drawback to these fabrics is that a cut fabric easily ravels in knitting and reverse directions. Jersey fabrics can be made in circular or flat knitting machines with a set of needles in circular or linear positions.

Rib structure differs from the jersey fabric in that it has identical appearance in both directions. The rib fabric is produced when stitches intermesh in opposite directions on a walewise basis. When opposite interlocking occurs in every other wale, the fabric is known as 1 × 1 rib. Similarly, when the interlocking occurs at every three wales in one direction to every two wales in the opposite direction, the fabric is termed as 3 × 2 rib. The rib fabric can be produced on a simple circular knitting or on a flat knitting machine using additional attachments.

Purl fabrics are produced on machines with needles that have hooks at both ends. Purl structures have one or more wales that contain both face and reverse loops.

2.5.2 WARP-KNITTED STRUCTURES

In warp knitting, each yarn is knitted by one needle. The needle bar that carries the needle moves sideways as well as up and down, so that the yarns are carried vertically and, to a limited extent, diagonally. This diagonal motion is needed to assure that the yarns interlace not only with the stitch directly below but also with stitches to the side. The fabric is formed by the intermeshing of parallel warp yarns that are fed from a warp beam. Here, the warp yarns move in a zigzag motion along the length of the fabric which results in a loop at every change of direction as individual yarn is intermeshed with neighboring yarns. Compared to weft-knit fabrics, warp-knit fabrics are flatter, closer, less elastic, and dimensionally more stable, as parallel rows of loops are interlocked in a zigzag pattern. They have a higher production rate and can be produced in a wider width. Warp-knit fabrics can be of different types of construction, i.e., tricot, Raschel, simplex, and milanese. Among these, tricot and Raschel are commonly used in industrial textiles.

In warp knitting, guide bars are used to guide sets of yarns to the needles. The pattern potential of the knitted fabric is controlled by these devices. Tricot fabrics are classified according to the number of guide bars used. Thus, one-bar tricot uses one guide bar, and two-bar tricot uses two guide bars for production. Tricot fabrics are known for softness, wrinkle resistance, and drapability. They possess higher bursting and tear strength.

Compared to tricot warp-knit fabrics, Raschel fabrics are generally coarse gauge. However, the machines used for producing Raschel fabrics are more versatile, and they have a very large pattern area for ornamentation. Typical products made from this fabric include dressware, laces, powernets, swimwear, curtain nets, etc.

The majority of tricot fabrics are knitted from smooth filament yarns in lightweight construction. On the other hand, the Raschel machines are capable of producing heavier fabrics using spun yarns. Patterns of some important knits are given in Figure 2.3.

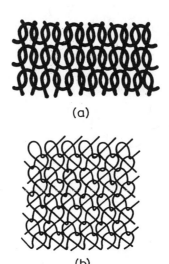

(a)

(b)

Figure 2.3 Patterns of some important knits: (a) jersey knit and (b) warp-knit tricot.

2.6 NONWOVEN FABRICS

Nonwoven fabrics are constructed directly from a web of fibers without the intermediate step of yarn manufacture as is necessary for woven, knitted, braided, or tufted fabric. These fabrics are extensively used in disposable and reusable goods because of their low cost and suitability for several specialized applications, as in fusible interlinings, filter media, surgical wear, sanitary goods, diapers, and wipers, etc. In the coating industry, they are widely used in synthetic leather, poromorics, upholstery backing, and protective clothing. Nonwovens may be classified by the type of fiber used, method of web formation, nature of bonding, and type of reinforcements used. The fibers commonly used are cotton, nylon, polyester, rayon, acetate, olefins, and combinations. There are two distinct steps in the manufacture of nonwovens. The first step is the manufacture of a web of fiber. The laid fibers, known also as a batt, do not possess adequate strength. The second step involves entanglement or bonding of the fibers to develop adequate strength.

2.6.1 WEB FORMATION

There are various methods for laying the web. In mechanical methods, compressed fibers are passed over rotating wire-covered cylinders (carding machine). The wires pick up the fibers and deposit them in sheet or batt form. A single layer of the web produced from the card is too thin, as such, multiple layers are often stacked to achieve the desired thickness. If the web layers are laid in parallel, they are known as parallel-laid web. These webs have higher strength in the machine direction than in the cross direction. If the webs obtained from the card are cross lapped (changing the orientation from direction of web travel to cross direction), the ratio of strength between the machine direction to cross direction is reduced. Such webs are known as cross-laid webs.

An effective way to minimize the fiber alignment is to sweep the opened fiber coming out of a carding machine by a stream of air and then condense the fiber on a slow moving screen or perforated drum. Such webs are known as air-laid webs. Webs can also be produced by a wet process similar to the one used in the making of paper. In this process, the fibers are suspended in water. The suspension is passed over a moving screen to remove the water. The remaining water is squeezed out of the web, and the web is dried. Webs produced by this method are denser than those produced by the air-laid process. The fiber alignment is also more random.

The spunbonded method is especially used for the manufacture of a nonwoven from continuous filament fibers. In this process, continuous filament extruded through spinnerettes is allowed to fall through a stream of air on a moving conveyor. The desired orientation of the filaments in the web is achieved by controlling the stream of air, speed of conveyor, and rotation of the spinnerette.

If the fibers are thermoplastic, the batt can be thermally bonded by passage between the nip of the heated calender roll. Fusion of the fibers occurs at the intersections.

2.6.2 WEB BONDING

There are two types of bonding for the batt: entanglement of fibers and bonding by adhesives.

2.6.2.1 Entanglement of Fibers

(1) Needle punching process: this is one of the most common mechanical bonding processes. In this process, an array of barbed needles is pushed through the web. The barbs hold the fibers at the surface and push them into the center, densifying the structure and leading to an increase in strength. The machine consists of a bed plate to support the web as the set of needles penetrates the web and a stripper plate to strip the fabric off the needles. The number of penetrations per unit area controls the density, thickness, and permeability of the nonwoven fabric. Most needle-punched fabrics are reinforced by a scrim.

(2) Hydroentanglement: in this process, a fine jet of water is used to push fibers from the surface toward the interior of the batt. During impingement, the web is supported on a bed. The force exerted by the jet is less than that

exerted by needling, and as such, the hydroentangled structure is less dense and more flexible than that obtained by the needling process. The method is useful for bonding relatively thin webs. Such fabrics are often termed spunlaced fabrics.

2.6.2.2 Bonding by Adhesives

The bonding of fibers of the web can be achieved by using a variety of adhesives, both in liquid and solid forms. Liquid adhesives can be solutions, emulsions, or pastes. They are applied on the web by dipping and squeezing, spraying, or kiss coating. After application of the adhesives, the solvent is evaporated, and the adhesive is cured by passage of the web through a heated chamber using heated air or IR heaters.

Solid adhesives can be applied as hot melt by spraying or by using the gravure/rotary printing process and subsequently cooling the web. Solid powder adhesives can be applied on the web by scatter coating; the adhesive is activated by passage through a heated chamber. If the web contains a mixture of low melting fibers along with high melting or nonmelting fibers, e.g., rayon and polyester, passage of the web through a heated chamber or hot calender rolls melts the thermoplastic fiber, leading to bonding. These are known as thermobonded fabrics.

The distribution of adhesive in the web is very important, because unlike in paper, movement of fibers in a nonwoven fabric is necessary to produce its textile-like properties. Adhesives interfere with fiber movement. Small bonded areas separated by unbound areas constitute the textile property of the fabric. If the adhesive fills the void between the fibers completely, the product becomes similar to fiber-reinforced plastic.

The strength of a web can be enhanced by the incorporation of yarn in the web. In the stitch-bonded process, the web is passed through a sewing or knitting machine. The stitched structure holds the fibers of the web together. Thus, the fabric is an open-mesh yarn structure with interstices filled with nonwoven fibers. The technology is known as the "Arachne" stitch bonding process.

2.7 REFERENCE

1. W. C. Smith, Journal of Coated Fabrics, vol. 15, Jan., 1986, pp. 180–197.

2.8 BIBLIOGRAPHY

S. Adanur, Ed., *Wellington Sears Handbook of Industrial Textiles,* Technomic Publishing Co., Inc., Lancaster, PA, 1995.

M. Grayson, Ed., *Encyclopedia of Textiles Fibres and Non Woven Fabrics,* John Wiley & Sons, New York, 1984.

A. J. Hall, *Standard Handbook of Textiles,* Newness-Butterworth, London, 1975.

F. Happey, Ed., *Contemporary Textile Engineering,* Academic Press, London, 1982.

M. Lavin, and Sello, S. B. Eds., *Handbook of Fibre Science and Technology, Chemical Processing of Fibres and Fabrics, Fundamentals and Preparation,* Part A, vol. 1, Marcel Dekker, New York, 1983.

H. F. Mark, Atlas S. M., and Cerina, E. Eds., *Man Made Fibres Science and Technology,* vols. 2 and 3, Interscience Publishers, New York, 1968.

R. Mark, and Robinson, A. T. C., *Principles of Weaving,* The Textile Institute, Manchester, 1976.

R. W. Moncrieff, *Man Made Fibres,* John Wiley & Sons, New York, 1963.

V. Mrstina and Fejgl, F., *Needle Punching Textile Technology,* Elsevier, New York, 1990.

A. T. Purdy, *Non Woven Textiles, Textile Progress,* vol. 12, no. 4, Textile Institute, U.K., 1983.

W. Scott-Taggart, *Cotton Spinning,* Universal Publishing Co., Bombay, 1985.

CHAPTER 3

Coating Methods

3.1 GENERAL FEATURES

COATING a layer of polymeric material on a textile imparts new character-istics to the base fabric. The resultant coated fabric may have functional properties, such as resistance to soiling, penetration of fluids, etc., or have an entirely different aesthetic appeal, such as finished leather. There are various coating methods used to apply polymer to textiles. They can be classified on the basis of equipment used, method of metering, and the form of the coating material. The various methods are given below.

(1) Fluid coating: the coating material is in the form of paste, solution, or latices.

   a. Knife coaters, wire wound bars, round bars, etc.: these are post-metering devices.

   b. Roll coaters, reverse roll coaters, kiss coaters, gravure coaters, dip coaters, etc.: these are premetered application systems.

   c. Impregnators: material to be coated is dipped in the fluid, and the excess is removed by squeeze roll or doctor blades.

   d. Spray coaters: the material is sprayed directly on the web or onto a roll for transfer.

(2) Coating with dry compound (solid powder or film):

   a. Melt coating: extrusion coating, powder coating, etc.

   b. Calendering: for thermoplastic polymers and rubber compounds, Zimmer process, etc.

   c. Lamination

The choice of a coating method depends on several factors. They are as follows:

75

- nature of the substrate
- form of the resin and viscosity of the coating fluid
- end product and accuracy of coating desired
- economics of the process

3.1.1 COMMON FEATURES OF FLUID COATING UNITS

A fluid coating operation basically involves applying the coating fluid onto the web and then solidifying the coating. There are common features in all coating operations. The different modular sections of a coating machine are illustrated in Figure 3.1 and described below.

(1) Fabric let-off arrangement. Here, the base fabric is unwound and drawn through the machine under uniform tension. Many machines have accumulator sections, where the rolls are temporarily sewn together for continuous operation, without interruption due to changeover of the rolls. A typical accumulator is shown in Figure 3.2.

(2) A coating head. It may be knife, roll, or any of the methods of fluid coating.

(3) Drying oven. All of the solvents are evaporated, and the film is solidified, dried, and cured. The oven may be steam heated, air heated (oil/forced air), or electrically heated. For rubber-coated fabrics, vulcanization is carried out separately after removal of solvents by evaporation. For other polymers requiring higher temperature, drying and curing can be done by IR heaters, gas-fired units, heater strips, etc. To prevent volatiles from forming an explosive mixture, fresh air is continuously circulated throughout the oven. In the case of organosols, the drying rate is carefully controlled to prevent blister formation or cracking. To properly control solvent evaporation, it is necessary to divide the oven into several zones, increasing the temperature of each zone in order to remove the solvent without blisters.

(4) Winding section. The fabric coming out of the oven is passed over cooling drums to make it tack free. The fabric is then wound up in rolls.
   In addition, there is a drive unit that transports the substrate web through the

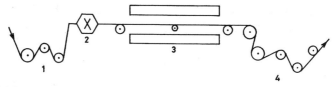

Figure 3.1 Layour of direct coating line: (1) fabric let-off arrangement, (2) coating head, (3) drying oven, and (4) winding section. (Adapted with permission from G. R. Lomax, *Textiles*, no. 2.1992, ©Shirley Institute U.K. [1].)

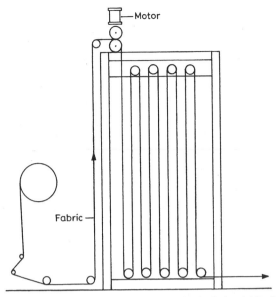

Figure 3.2 Line diagram of a fabric accumulator. Courtesy Sanjay Industrial Engineers, Mumbai, India.

coating head under constant tension. At times, the drive unit incorporates a stenter frame to minimize shrinkage during the drying process. Coating thickness can be measured by a β-ray gauge or from the web speed and flow rate of the coating fluid [1,2]. A general view of a coating plant is shown in Figure 3.3.

Figure 3.3 General view of a fluid coating plant. Courtesy Polytype, U.S.A.

3.1.2 POST- AND PREMETERING METHODS

The coating process can be classified on the basis of stages of metering, i.e., (a) process where the material is applied on the substrate and then metered and (b) process where the material is metered prior to application. A combination of these methods uses application of premetered excess followed by further metering for more accurate coating. A discussion of the advantages and disadvantages of the methods is presented below.

The first category, i.e., postmetering processes, is considered to be effective for coating noncritical weights on the substrate. As has been described earlier, in this class are the common knife coaters, wire wound (Mayer rod) coaters, single-roll squeeze coaters, etc. Here, excess coating is initially applied on the textile substrate. After the substrate is wetted, a coating device meters the coating to a predetermined thickness. The parameters necessary for consistency in coating add-on are as follows:

a. Substrate tension

b. Viscosity of the coating material

c. Substrate uniformity and porosity

Any variation in these parameters may lead to a nonuniform coating. Coating accuracy is poor. The coating range is limited to about 0.02 to 0.2 mm thickness. However, the major advantage is their low investment cost and fast product changeover.

In the second category, a premetered quantity of material is applied onto the textile. The processes include roller coatings, gravure coatings, extrusion coatings, and lamination. These methods are much more accurate and give highly reproducible add on. The coating range is wider, 0.1 to 0.5 mm. However, the initial investment cost is higher [3].

The common methods of coating are described below.

3.2 KNIFE COATING

Also known as spread coating, this is one of the oldest coating methods. A dry, smooth fabric is fed over the bearer roll under a knife known as a knife or doctor blade. The coating material is poured in front of the knife by a ladle or by a pump over the entire width of the web. As the web is transported under the knife, the forward motion of the fabric and the fixed knife barrier give the viscous mass of the material a rotatory motion. This is known as the rolling bank that functions as a reservoir of coating compound in front of the knife. To prevent the fluid from spilling over the edges of the fabric, two adjustable guard plates known as dams are also provided. Proper tension is applied on the fabric as it is unwound, so that the fabric is taut under the knife. Most machines can coat fabric widths up to 1.5–2.0 m, but specially designed machines can accommodate up to 4 m widths. Special care is taken that the material has adequate viscosity so that it does not strike through the fabric. The coated fabric then passes through the drying oven. The rate of evaporation of the solvent determines the rate of transport of the fabric, and thus, the coating rate. The coating thickness is mainly controlled by the gap between the knife and the web [4].

3.2.1 ARRANGEMENTS OF KNIFE COATING

There are three distinct arrangements of knife coating. They are knife on air, knife on blanket, and knife on roll. These arrangements are given in Figure 3.4.
In the floating knife [Figure 3.4(a)] or knife-on-air coating, the knife is positioned after a support table and rests directly on the fabric. In this arrangement, compressive force applied on the coating material is greater, and as such, the coating compound enters the interstices of the fabric. This technique is useful for applying very thin, lightweight, impermeable coatings (as low as 7–8 g/m^2) suitable for hot air balloons, anoraks, etc. [5].

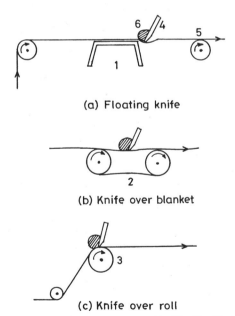

(a) Floating knife

(b) Knife over blanket

(c) Knife over roll

Figure 3.4 Different types of knife coating: (1) support table, (2) rubber blanket, (3) rubber or steel roll, (4) knife, (5) web, and (6) coating material. (Adapted with permission from *Encyclopedia of Chemical Technology*, Vol. 6, 3rd Ed. 1979; and *Encyclopedia of Polymer Science & Engineering*, Vol. 3, 2nd Ed. 1985. Both ©John Wiley & Sons.)

Web tension, viscosity, percent solids, and specific gravity of the coating compound play a significant role in the amount of coating deposited. The higher the viscosity of the compound, the greater will be its tendency to force the web away from the knife, resulting in a higher weight add on. On the other hand, it can be easily visualized that the coating weight will be less if tension on the fabric is greater. The method is suitable for both closely woven and open fabrics, because strike through does not affect the coating operation [6].

In the knife-on-blanket arrangement [Figure 3.4(b)], the web is supported by a short conveyor, in the form of an endless rubber blanket stretched between two rollers. Because the tension applied on the blanket results in a uniform pressure between the knife and the substrate, the fabric is not subjected to stretching in this arrangement. It is possible to coat dimensionally unstable substrates with this technique. The amount of coating is dependent on the tension of the blanket, which is adjusted by the rollers. Care should be taken that there is no damage to the blanket and that no foreign matter is adhered on the inside of the belt, as this will result in an irregularity in coating weight.

The knife-on-roll system [Figure 3.4(c)] is the most important and widely used technique for its simplicity and much higher accuracy. In this configuration, a suitably designed doctor blade is properly positioned on top of a high-precision roller. The gap between the bottom of the blade and the thickness of the fabric that passes over the roller controls primarily the coating weight. The roll may be rubber covered or chromium-plated steel roll. The hardness of the rubber-covered roll may vary from 60 to 90 shore A, depending upon the type of fabric [4]. The advantage of a rubber-covered roll is that any fabric defects, such as knots and slubs having thickness greater than the fabric thickness, are absorbed by the roll surface, allowing free passage of the fabric through the coating knife. However, rubber rolls are not as precise as steel rolls and may cause variation in the wet coating weight up to ±30 g/m^2. Rubber rolls also have the disadvantage of swelling on prolonged contact with solvents and plasticizers. Steel rolls, on the other hand, can give more precise coating [5]. The gap between the knife and the roll can be adjusted by the screws provided on the mounting rod of the knife. In modern machines, the gap is controlled by pneumatic arrangement. Besides pneumatic, quick lifting is also provided to release lumps, splices, torn edges, etc. Materials with a wide viscosity range (up to 40,000 cps) can be coated by the knife-on-roll technique. It can also impart heavy coatings on fabric using a solventless system like plastisols. The coating method is better suited for dimensionally stable fabrics, which will not easily distort due to tension applied on the fabric while pulling through the coater. Care is always taken that the coating material does not strike through. In case of a strike through, steel rolls are easier to clean. A knife-on-roll coating plant is shown in Figure 3.5.

Figure 3.5 A knife on roll coating plant. Courtesy Egan Davis Standard Corp., and U.S.A.

3.2.2 COATING KNIVES

The profile of the coating knife and its positioning over the roll are important parameters affecting coating weight and penetration. Numerous knife profiles are used in the trade, however, some common types are shown in Figure 3.6.

The knife profile [Figure 3.6(a)] is normally used for lightweight coating. The base of the knife may vary from 0.5 to 4 mm wide. The knife is chamfered on the other side of the rolling bank. The sharper the base of the knife, the lower the coating weight. If the blade is chamfered on both sides, that is the V-type profile [Figure 3.6(b)] a wedge effect is produced during coating,

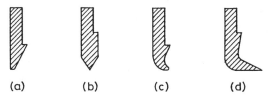

(a)   (b)   (c)   (d)

Figure 3.6 Profiles of knives: (a) knife type, (b) V type, (c) bull nose, and (d) shoe. (Adapted with permission from F. A. Woodruff. *J. Coated Fabrics*, Vol. 21, April 1992. ©Technomic Publishing Co., Inc. [5].)

which puts considerable pressure on the coating material, resulting in much greater penetration of the material into the interstices of the fabric. This type of profile is used where a high degree of penetration is required for good mechanical adhesion. Multiple coats are applied to achieve the desired coating weight of the end product, such as for tarpaulins, hoses, etc. A bull-nosed knife [Figure 3.6(c)] imparts heavy coating weights with little penetration into the weave and is suitable for easily damaged fabrics. The shoe [Figure 3.6(d)] knife is so named because of its resemblance to a shoe. The front of the knife may be straight or rounded. The base dimension may vary from 2–30 mm. The toe of the blade is nearest to the substrate. By varying the angle between the blade and the roll, the elevation of the heel of the knife and the web can be altered. A wedge of coating compound is formed between the web and the heel of the knife. The greater the elevation, the more material will be available in the wedge, leading to greater penetration [5].

During the coating process, many compounds, because of their surface tension properties, rise up the back of a knife, accumulate, and drop on the coated surface in an unsightly fashion, called spitting. In the shoe knife, due to the design of the toe, this is completely prevented. For PVC pastes and breathable PU coatings, shoe knife is the preferred type [6].

Proper positioning of the knife and the roll are other important considerations for proper coating. The rolls should be true to knife surface, without eccentricity. It is also vital that the knife be aligned horizontal to the axis of the roll, otherwise, wedge-shaped coating will result. The angle of the knife over the roll affects penetration. The greater the angle at which the knife meets the moving fabric, the greater the penetration. If the position of the blade is at a point behind the crown of the roll, the blade will be directly pressing the fabric, and a situation similar to floating knife is created [5].

Instead of a single knife fitted over the roll, modern machines have a knife supporting beam fitted with two or three different types of knives mounted 180° or 120° apart. A twin knife arrangement is shown in Figure 3.7. This facilitates easy changeover of the blades.

3.2.3 THE ROLE OF ROLLING BANK AND WEB TENSION

As has been described earlier, the coating compound is pumped or ladled over the substrate in front of the blade in knife coating, forming a rolling bank that acts as a reservoir for the coating material. The viscosity and the amount of rolling bank also contribute to the penetration and coating weight of the end product. The rolling bank (Figure 3.8) exerts a pressure on the web, as such, if the height of the roll bank is greater at the center, heavier coating will be produced at the center. Similarly, if the height is more at the sides, the coating will be more in the sides. In case of a pump pouring fluid over the web across the width, its traverse also results in variation of coating in an "S" pattern.

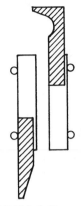

Figure 3.7 Twin knife arrangement.

Harrera has described Mascoe's patented trough system (Figure 3.9) for better control of the conditions prevailing in the rolling bank [7]. The coating compound is fed into a trough in front of the blade. The gap between the trough and the blade is adjustable, but it is fixed during the coating process.

In this device, because the opening of the trough is constant, the exposure to fabric is controllable and is much shorter than the rolling bank. It is claimed that greater accuracy and repeatability can be obtained by this system, however, cleanup of the trough is a problem.

Fabric tension plays an important role in the final add on of the coated product and is thus dependent on the stretchability of the fabric. A higher tension in warp direction opens up the weave, exposing more surface, and thus, the coating weight add on is heavier when the fabric is excessively stretched. This is true if the weave pattern is regular. In case of an irregular pattern, application of uniaxial tension results in uneven tension in the filling yarn, causing uneven coatings. Mascoe has developed a tensioner that maintains a uniform force across the substrate width, which is not altered by the changing speed of the web [7].

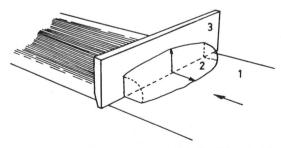

Figure 3.8 Rolling bank: (1) web, (2) rolling bank, and (3) knife. (Adapted with permission from A. Harrera. *Journal of Coated Fabrics*, Vol. 20, April 1991. ©Technomic Publishing Co., Inc. [7].)

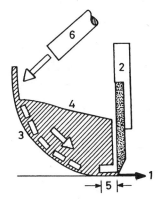

Figure 3.9 Mascoe's trough: (1) web, (2) knife, (3) trough, (4) coating compound, (5) adjustable gap, and (6) feeder. (Adapted with permission from A. Harrera. *Journal of Coated Fabrics,* Vol. 20, April 1991. ©Technomic Publishing Co., Inc. [7].)

3.3 ROLL COATING

3.3.1 MAYER ROD COATING

In this method, compound is applied on the web by a single-roll applicator. The coating is postmetered by a wire wound rod, known as the Mayer rod, that removes excess coating (Figure 3.10).

The Mayer rod is a small, round stainless rod, wound tightly with a fine wire also made of stainless steel. The grooves between the wire determine the precise amount of coating that will pass through. The coating thickness is directly proportional to the diameter of the wire. The most common core rod diameter varies from 4–6 mm, although sizes up to 25 mm are used. To prevent deflection of the Mayer rod due to web pressure, it is mounted on a rod holder. The simplest rod holder is a rectangular steel bar with a "V" groove machined to it. The rod is placed on the groove, and the holder is mounted between the side frames of the coating machine. During coating, the rod is slowly rotated

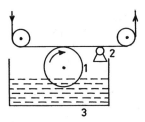

Figure 3.10 Mayer rod coater: (1) applicator roll, (2) Mayer rod with holder, and (3) feed pan. (Adapted with permission from *Encyclopedia of Polymer Science & Engineering,* Vol. 3, 2nd Ed. 1985, ©John Wiley & Sons.)

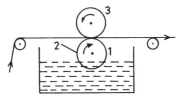

Figure 3.11 Direct roll coater: (1) applicator roll, (2) doctor blade, and (3) backup roll.

in the opposite direction of the web. The rotation removes the coating material between the wires, keeping the wire surface wet and clean. The rotation also increases the life of the rod due to reduced wear.

The thin lines formed during coating smooth out due to surface tension. The uniformity of the coating is maintained if the viscosity of the compound, speed, and tension of the web are properly controlled. This method is used for low solid, low viscosity (50–500 cps), thin coatings (2–3 g/m^2). It is suitable for silicone release papers and as a precoater [8].

3.3.2 DIRECT ROLL COATING

In direct roll (or squeeze roll) coating, a premetered quantity of the coating is applied on the fabric by controlling the quantity on the applicator roll by the doctor knife (see Figure 3.11.) The fabric moves in the same direction as the applicator roll. This method is also restricted to low viscosity compounds and is suitable for coating the undersurface of the fabric. The coating thickness depends on nip pressure, coating formulation, and absorbency of the web [2].

3.3.3 KISS COATING

A typical arrangement of kiss coating is shown in Figure 3.12.

The pickup roll picks up coating material from the pan and is premetered by the applicator roll. The coating is applied on the web as it kisses the applicator roll. The pickup roll may be rubber covered, and the applicator roll may be made of steel. The metering is done by nip pressure, and consequently, the

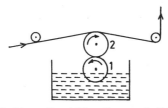

Figure 3.12 Kiss coater: (1) pickup roll and (2) applicator roll. (Adapted with permission from *Encyclopedia of Polymer Science & Engineering*, Vol. 3, 2nd Ed. 1985, ©John Wiley & Sons.)

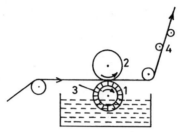

Figure 3.13 Gravure coater: (1) gravure roll, (2) backup roll, (3) doctor blade, and (4) smoothening rolls. (Adapted with permission from *Encyclopedia of Polymer Science & Engineering*, Vol. 3, 2nd Ed. 1985, ©John Wiley & Sons.)

amount of material coated on the web is dependent on nip pressure, speed of the operation, roll hardness, and its finish. The coating weight and splitting of the film as it leaves the roll are also dependent on web tension.

3.3.4 GRAVURE COATING

Engraved rollers are utilized in gravure coatings to meter a precise amount of coating on the substrate. The coating weight is usually controlled by the etched pattern and its fineness on the gravure roll. There are a few standard patterns like the pyramid, quadrangular, and helical. For lighter coating weight, a pyramid pattern is used. In a direct two-roll gravure coater (Figure 3.13), the coating material is picked up by the gravure roll and then transferred to the web as it passes between the nip of the gravure and the backup roll. The pattern may be self-leveling or the coated web may be passed between smoothening rolls.

In offset or indirect gravure coater, a steel backup roll is added above the direct gravure arrangement. The coating compound is first transferred onto an offset roll and then onto the substrate (Figure 3.14).

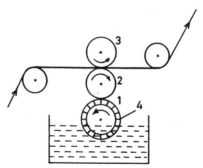

Figure 3.14 Offset gravure coater: (1) gravure roll, (2) rubber-covered offset roll, (3) steel backup roll, and (4) doctor blade. (Adapted with permission from *Encyclopedia of Polymer Science & Engineering*, Vol. 3, 2nd Ed. 1985, ©John Wiley & Sons.)

Figure 3.15 A gravure coating plant. Courtesy Polytype, U.S.A.

The speed and direction of the gravure and offset rollers can be varied independently. The arrangement is suitable for an extremely light coating (as low as 0.02 g/m^2) and minimizes the coating pattern. This offset process can handle a higher viscosity material ($\sim$10, 000 cps.) than the direct process. By heating the feed pan, the process can coat hot melt compounds. Gravure coating is used for applying laminating adhesives or a topcoat on a treated fabric. A gravure coating plant is shown in Figure 3.15.

3.3.5 REVERSE ROLL COATERS

Reverse roll coating is one of the most versatile and important coating methods. It can be used for a wide range of viscosities and coating weights. The accuracy of the coating is very high. Reverse roll coaters apply a premetered coating of uniform thickness, regardless of the variations in substrate thickness, and are therefore known as contour coaters. The coating is also independent of substrate tension. There are two basic forms of reverse roll coaters: three-roll nip and pan fed. Figure 3.16 shows the arrangement of a nip-fed coater.

The applicator and the metering rolls are precision-ground chilled cast iron or stainless steel rolls, finished to a high degree of precision. The two rolls are set at an angle, and the coating material is kept in a reservoir, at the nip, bound by the applicator roll and coating dams on each side. The gap between the applicator and metering roll can be precisely controlled. The backup roll is to bring the moving web in contact with the applicator or transfer roll. A film

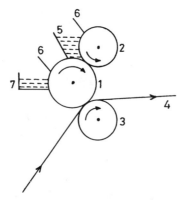

Figure 3.16 Nip-fed reverse roll coater: (1) applicator roll, (2) metering roll, (3) backup rubber roll, (4) web, (5) coating pan, (6) doctor blades, and (7) drip pan. (Adapted with permission from *Encyclopedia of Polymer Science & Engineering*, Vol. 3, 2nd Ed. 1985, ©John Wiley & Sons.)

of the coating compound is metered between the applicator and the metering roll. The applicator roll then carries the coating material to the coating nip where the compound is transferred to the web moving in the opposite direction. The opposite direction of the applicator roll and the web creates a high level of shearing action. The criteria of a reverse roll coater are the opposite direction (a) of the applicator and the metering roll and (b) of the applicator roll and the web. A scraper or doctor blade cleans the metering roll to prevent dropping of material on the web. The coating material remaining on the applicator roll after contact with the web is also scraped off, collected in a pan, and recycled. This helps to clean the roll of dirt and dried coating material, which would cause inaccuracy in coating.

The thickness of the coating is controlled by the gap between the applicator and the metering roll, the rotational speed of the applicator roll, and the amount of material transferred on the web, which in turn is dependent on the web pressure on the applicator roll adjusted by the backup roll. Thus, a reverse roll coater has greater flexibility in adjusting the coating thickness compared to the knife-on-roll, where coating is controlled only by gap of knife and the roll. In direct roll coating, where the web and applicator roll move in the same direction, nonuniform coating occurs with the formation of ribbing due to the film split phenomenon, while in reverse roll, the coating is smooth [6].

One challenge when using the nip-fed coater is to prevent leaks from the coating reservoir, particularly with low viscosity compounds. The pan-fed coater operates using the same principle as the nip-fed coater, but it is more suited for low viscosity materials (Figure 3.17).

Greer [9] has recently reviewed studies on the fluid mechanics of reverse roll coating. The reasons for nonuniformity of coating have been discussed. The two common defects are ribbing and cascading, i.e., formation of a wavy pattern. To

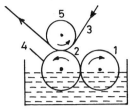

Figure 3.17 Pan-fed reverse roll coater: (1) metering roll, (2) applicator roll, (3) web, (4) doctor blade, and (5) backup roll. (Adapted with permission from *Encyclopedia of Polymer Science & Engineering*, Vol. 3, 2nd Ed. 1985, ©John Wiley & Sons.)

explain the reasons for these defects, a dynamic wetting line has been defined. This is the point from which the coating pulls away from the metering roll as it leaves the metering nip and moves along with the applicator roll. If the wetting line is at the center of the nip, the coating is smooth. If the wetting line is at the outlet side, ribbing occurs; if it is on the inlet side, cascading occurs. An important criterion governing the position of the wetting line is the ratio of metering to applicator roll speeds. In addition, surface tension and viscosity of the coating compound also play important roles.

3.4 DIP COATING

This is also known as impregnation or saturation. The substrate web is immersed in a tank of the coating material for a certain period of time, known as the dwell time. The excess material is then squeezed out by passing through nip rolls or a set of flexible doctor blades precalibrated to give a fixed net pickup of the resin. There are various arrangements for the dip process. A simple arrangement is shown in Figure 3.18. Sometimes, a prewet station precedes the dipping to remove air from the interstices and promote penetration. The factors that are to be considered in designing an impregnated fabric are the solid content of the impregnant and the absorption capacity of the fabric. In dip coating, the pickup is quite low, and penetration occurs into the interstices of the fabrics as well as in the yarns. Moreover, because the fabric is not stressed,

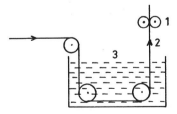

Figure 3.18 Dip coating: (1) squeeze rolls, (2) web, and (3) dipping tank. (Adapted with permission from *Encyclopedia of Polymer Science & Engineering*, Vol. 3, 2nd Ed. 1985, ©John Wiley & Sons.)

no damage or distortion to the yarn occurs. The process is mainly used for finishing processes like flame retardant treatments and application of adhesive primer.

3.5 TRANSFER COATING

In principle, transfer coating consists of applying polymeric coating on the surface of a support, usually paper, laminating the textile substrate to be coated to the polymeric layer, and removing the paper, to yield a transferred polymeric layer on the textile.

The process of applying coating material directly on the textile is known as direct coating. The direct coating process has certain limitations. They are as follows:

- It is applicable to closely woven, dimensionally stable fabrics that can withstand machine tension, and it is not suitable for excessively stretchable knitted fabrics.
- Penetration occurs in the weave of the fabric, increasing adhesion and lowering tear strength and elongation, resulting in a stiff fabric.

Transfer coating overcomes these limitations. Because no tension is applied during coating, the most delicate and stretchable fabrics can be coated by this process. Fabric penetration and stiffening is significantly low. Moreover, with proper processing, the appearance of the textile substrate can be altered to give a much better aesthetic appeal, like artificial leather for fashion footwear. A schematic diagram of the process is given in Figure 3.19.

The steps involved are as follows [1]:

- A layer of coating is applied on a release paper in the first coating head and is then passed through the first oven, where it is dried and cured. This forms the top surface of the coated fabric. The release paper is

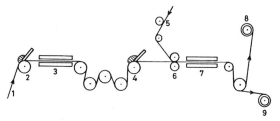

Figure 3.19 Layout of transfer coating process: (1) release paper, (2) first coating head, (3) first oven, (4) second coating head, (5) textile substrate, (6) laminating nip rolls, (7) second drying oven, (8) coated fabric takeoff roll, and (9) release paper wind roll. (Adapted with permission from G. R. Lomax, *Textiles*, no. 2. 1992. ©Shirley Institute U.K. [1].)

usually embossed. The pattern of the paper is thus transferred on the coating. This is known as the top coat.

- In the second coating head, an adhesive layer known as the tie coat is applied on the dry top coat, previously laid on the release paper. The release paper thus has two layers, the dry top coat and the tacky adhesive tie coat.
- The textile substrate is then adhered to the release paper containing the top and the tie coats, while the tie coat is still tacky. The lamination is done by a set of nip rolls. The composite layer is then passed through the second oven to dry and cure the tie coat.
- The release paper is finally stripped, leaving the coated textile.

The release papers are calendered to uniform thickness and are coated with a thin layer of silicone. The property of release paper should be such that it is able to grip the top coat during the processing and able to release the fabric without damaging the top coat. Various grades are available. Normally, paper can be reused about eight to ten times.

The coating head in a transfer coating unit is typically knife over rubber-backing roll. The rubber roll has the advantage in that it does not damage the release paper. The laminator rolls are steel rolls. The setting of the coating knives and gap between the laminating rolls can be set and maintained fully automatically.

Transfer coating is used for PVC pastes and for polyurethane coating. Although the basic transfer process involves a two-coat operation, the top and the tie coat, a three-coat process is becoming quite popular. The first two heads apply the top coat in two thin layers. This permits faster line speeds due to greater efficiency of solvent removal from thinner films, and it prevents pin-holing, where waterproofness is important. The third coating head applies the tie coat. In polyurethane transfer coating, this affords an option of using two different types of PU for the two layers of top coats for special properties, as required for artificial leather [10].

3.6 ROTARY SCREEN PRINTING

This method is common for coating and printing textiles. The coating head is a screen that is a seamless nickel cylinder with perforations. This screen rests on the web. A squeegee is mounted in the screen, serving as supply and distribution pipe of the coating paste. The squeegee blade, which is mounted to this pipe, pushes the paste out through the perforations of the screen. A whisper blade smooths the applied coating. A backup roll is provided for counterpressure (Figure 3.20). After coating, the coated material is sent to an oven for fusion of the polymers. The amount of coating applied is determined mainly by the

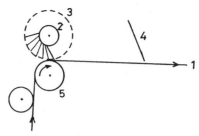

Figure 3.20 Rotary screen coating: (1) web, (2) squeegee, (3) screen, (4) whisper blade, and (5) backup roll.

mesh number of the screen, the squeegee pressure, i.e., the angle between the blade and the screen, and the viscosity of the paste.

Depending on the mesh size and design of the screen, continuous coating, coating of complex pattern, and dot coating can be done. In continuous coating, the coating can be up to 200 g/m^2, by proper choice of the screen. Dot coating is useful for making fusible interlinings for woven and nonwoven fabrics. In this process, the screen, the web, and the counterpressure roller all have same speed. The coating is, therefore, done without tension and friction. Consequently, delicate and stretchable fabrics can be coated without difficulty. The coating is accurate, and the penetration can be controlled [8].

A relatively new development by Stork (Stork-Brabent, Holland) is the screen-to-screen technology (STS). Basically, the process consists of two screen-coating heads, back to back, each with its own coating feed system, squeegee roll, and whisper blade to smooth out the applied compound. The substrate travels between the screens either in a horizontal or a vertical position (depending on the model), and the compound is gelled (or cured) with IR heaters. With STS technology, it is possible to coat (different colors) or print both sides of a substrate in one pass.

3.7 CALENDERING

Calendering is a versatile and precise method of coating and laminating polymeric material onto a fabric. The equipment consists of a set of heated rolls also known as bowls. Fluxed, precompounded stock is fed between the roll nips, which comes out as a sheet as it passes through consecutive roll nips. The sheet so produced is press laminated to the fabric with another pair of mating rolls, which may be in the same calender machine. A variety of thermoplastics can be processed on the calender, however, it is extensively used for coating rubbers and vinyls.

For continuous operation of a calender, different equipment is arranged in line and functions in tandem to produce the coated fabric. This is known as

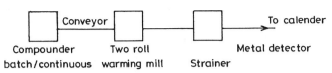

Figure 3.21 Precalender train.

the calender train. The train has three distinct sections, viz., the precalendering section, the calendering itself, and the postcalendering section.

3.7.1 PRECALENDERING SECTION

The task of the precalender section is to deliver fused polymer stock to the calender in a thoroughly compounded, homogenized, degassed condition that is free from impurities. The compounding can be done either in a batch process, using an internal mixer such as the Banbury mixer, or continuously, in an extruder. The compounded material is then fed to a two-roll mill to impart uniformity of temperature to the stock. Frequently, the material is next passed through a strainer. The strainer is a heated short-barreled extruder with a screen orifice. It masticates the compound further and prevents impurities, particularly metal particles, from entering the calender. The compounded material is then fed into the calender by a conveyor. The feed is a thick and narrow strip of material. Metal detectors are provided in the feed line to prevent any metal particle from entering the calender and damaging the rolls. If the material is conveyed a distance of more than 2 m, the feed is heated by IR heaters. The stages of this section are shown in the block diagram given in Figure 3.21.

3.7.2 CALENDER

The equipment consists of a stack of rolls mounted on bearing blocks, supported by a side frame. It is equipped with roll drive, nip adjusting gear, and heating arrangement. The rolls are made of chilled cast iron. The number of rolls and their arrangements are varied. Three- and four-roll calenders are the most popular.

Thinner coating can be achieved by increasing the number of rolls, but they increase the complexity of the equipment, the cost, and the space required. The sizes of the rolls vary from 45–120 cm diameter and 90–300 cm width [6,11,12].

Some common configurations of the rolls are given in Figure 3.22.

Vertical arrangements of the rolls in the stack were used in early machines. They suffer from the problem of adjusting the nips independently and of feeding the calender [13]. The feeding is easier if the top roll is offset, in an inverted "L"-type configuration, for instance, the feed bank is horizontal.

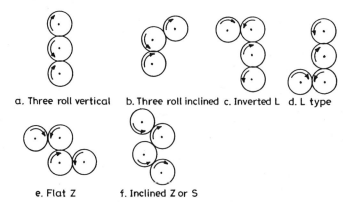

a. Three roll vertical   b. Three roll inclined   c. Inverted L   d. L type

e. Flat Z   f. Inclined Z or S

Figure 3.22 Configurations of rolls.

The selection of the configuration of rolls is dependent on its end use. The most suitable configuration for plasticized vinyl compounds is the inverted "L", the most suitable for rubbers is the three-roll inclined, and the most suitable for two-sided coating is the "Z" type [14].

The rolls are individually driven, which provides wide flexibility in the variation of the roll speeds and the corresponding friction ratio. For proper control of process temperature, the rolls are heated. The rolls are provided with either a hollow chamber or they have peripheral holes located close to the roll surface. The heating is done by circulation of hot water or special heat exchange liquid. Temperature control in the peripheral holes is far superior due to better heat transfer. The two factors, viz., the friction ratio and temperature, enable the calender to process a wide range of compositions differing in rheological properties. End dams are fitted in the rolls to constrain the compound in an adjustable span for coating fabrics of different widths [6,12].

The compound is fed into the feed nip of the calender. In the case of an inverted "L" type, it is the nip between the top two rolls. The rolls rotate in opposite directions at the nip and at different speeds. The material is pushed forward by the friction of the rolls and adheres to the faster roll. The material then passes through the successive rolls where it is resurfaced and metered and comes out of the calender in the form of a sheet. Calendering can, therefore, be regarded as sheet extrusion. During operation, rolling banks are formed at each nip. It is thick and narrow at the feed nip but becomes thin and wide in successive nips. The passage of the material from the feed to outlet is known as the sheet path and is controlled by the adherence of the material to a particular roll. The material adheres to the faster roll and the one having a higher temperature.

Separating forces on the rolls are produced when the viscous material is passed through the nip. These forces are highest at the center of the roll, therefore, the sheet is thicker in the middle. The separating force is dependent on several factors: the viscosity of the material, the gap between the rolls, the speed of the rolls, the size of the rolling bank, etc. In older machines, this was corrected by contouring the rolls in such a way that the center of the rolls had a slightly larger diameter, known as roll crowning. However, this is only suitable for specific compounds and operating conditions [15]. The methods adopted at present are roll bending and roll crossing. In roll bending, a hydraulic load is applied to the roll journal ends. This force exerts leverage on the rolls that makes them slightly concave or convex depending on the direction of the applied load. In roll crossing, an angular shift is given to one or both rolls at the nip. Although the rolls remain in the horizontal plane, their axes are no longer parallel but form a slight angle. This increases the end clearance between the rolls, resulting in thickening at the edges of the sheet produced. For proper operation, the material should be fed at a steady rate at uniform temperature. Any variation results in roll deflection and consequent variation in the thickness of the product.

3.7.3 COATING AND LAMINATION

A calender is used for coating polymer directly onto the fabric or for making unsupported film that may be subsequently laminated to a fabric. There are various ways of coating and lamination which are discussed below [6,11].

a. Nip coating: here, the coating is done at the bottom nip of the calender. The mechanical setup is simpler. The fabric pulls the sheet off the calender. The method is suitable for heavy impregnation. The extent of penetration is dependent on the gap at the nip and the friction ratio (Figure 3.23).

b. Lamination against calender roll: in this setup, the fabric is laminated against the last calender roll by means of a rubber backup laminating roll that is hydraulically operated—known as a squeeze roll. The conditions of penetration and takeoff of the sheet from the calender are similar to those in nip lamination (Figure 3.24).

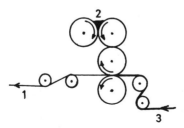

Figure 3.23 Nip coating: (1) fabric, (2) rubber bank, and (3) coated fabric. (Adapted with permission from *PVC Plastics* by W. V. Titow. ©Kluwer Academic Publishers, Netherlands, 1990.)

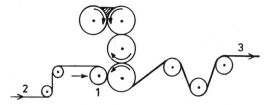

Figure 3.24 Lamination against calender roll: (1) squeeze roll, (2) fabric, and (3) coated fabric. (Adapted with permission from *PVC Plastics* by W. V. Titow. ©Kluwer Academic Publishers, Netherlands, 1990.)

c. In-line lamination: the sheet produced from the calender is laminated to the fabric outside the calender by laminating rolls. This arrangement is convenient for heat-sensitive substrates. The sheet coming out of the calender may cool and have to be heated prior to lamination for proper bonding to the fabric (Figure 3.25).

d. Another method of laminating multiple sheets of polymer and textile is given in Figure 3.26. Here, two or more sheets of polymer and textile are laminated by pressing them between a steel belt and a hot roll. The heat and pressure laminate the webs. The steel belt (1) is pressed against a hot drum (2) by means of a tension roll (3) and guide rolls (4 and 5). The fabric and the sheets are heated by IR heaters prior to being fed between the gap of the steel belt and the hot roll. The configuration is similar to the rotocure system used in continuous vulcanization.

e. Coating of elastomers: as mentioned earlier, the rolls rotate in the opposite direction at the nip with different speeds. The higher the friction ratio, the greater the penetration. Thus, if the rolls run at even or near even speeds, the penetration is low, and the coating thickness is high. For rubberized fabrics requiring thick coatings with high degrees of penetration for better adhesion, a friction coating is applied first, followed by top or skim coating. The frictioning is done at a higher temperature and at a friction ratio of

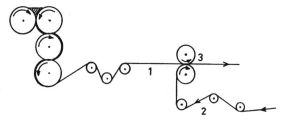

Figure 3.25 In-line lamination: (1) sheet, (2) fabric, and (3) hydraulically operated laminating roll. (Adapted with permission from *PVC Plastics* by W. V. Titow. ©Kluwer Academic Publishers, Netherlands, 1990.)

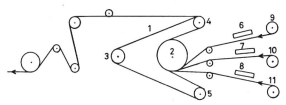

Figure 3.26 Lamination against steel belt: (1) steel belt, (2) hot roll, (3) tension roll, (4, 5) guide rolls, (6, 7, 8) IR heaters, (9, 11) polymer sheets, and (10) textile sheet. (Adapted with permission from D. Zickler. *Journal of Coated Fabrics*, Vol. 8, Oct. 1978. ©Technomic Publishing Co., Inc. [16].)

1:1.5 to 1:2; for skim coating, the friction ratio is 1:1.1 to 1:2. The operating temperature of the calender depends on the polymer, however, it is generally between 60° to 150°C. For rubber coating, the temperature required is lower to prevent scorching. A three-roll inclined calender is suitable for rubber coating (Figure 3.27) [14].

Simultaneous coating on both sides in a "Z"- type inclined calender is shown in Figure 3.28 [14].

3.7.4 POST-CALENDERING SECTION

A block diagram of the post-calender section is given in Figure 3.29.

Embossing consists of a pair of rolls, one of them is a metal engraved roll to impart the pattern, and the other is a rubber-covered roll. The coated fabric from the calender is passed through the nip of the rolls for embossing. The diameter of the roll is determined by the size of the repeat pattern. The rubber roll may have the same diameter or be larger than the metal roll.

The thickness of the sheet is measured by β-ray gauge. The feed from the gauge is used to automatically control roll bending or crossing to correct the thickness variation.

The finished product is then passed over cooling cans to lower the temperature and reduce tack. The train consists of a set of cans that is cooled by circulating

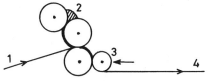

Figure 3.27 Three-roll inclined calender for rubberized fabric: (1) fabric, (2) rubber bank, (3) laminating roll, and (4) coated fabric. (Adapted with permission from J. I. Nutter. *Journal of Coated Fabrics*, Vol. 20, April 1991. ©Technomic Publishing Co., Inc. [14].)

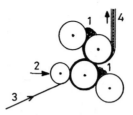

Figure 3.28 Coating on both sides in a "Z" calender: (1) rubber bank, (2) laminating roll, (3) fabric, and (4) double-sided coated fabric. (Adapted with permission from J. I. Nutter. *Journal of Coated Fabrics*, Vol. 20, April 1991. ©Technomic Publishing Co., Inc. [14].)

cold water through its inner shell. The number of cans depends on the sheet thickness and the speed of operation [6,11].

The fabric is then wound up in rolls.

3.7.5 COATING DEFECTS

One of the major problems of processing vinyl compounds is plate-out. It is the transfer of a sticky deposit that sometimes appears on the rolls of the calender. The calender rolls, embossing rolls, and cooling rolls require frequent cleaning in order to prevent loss of output and surface defects. Plate-out occurs from hot PVC stocks due to an incompatibility of some the constituents of the compound, particularly certain lubricants and stabilizers.

Other surface defects, like discoloration, surface roughness, crowfeet marks, bank marks, etc., are caused by diverse factors, such as excessive heat on the compound, nonuniformity of the stock temperature, improper dispersion of particulate additives, and undergelation of PVC [11,15].

3.7.6 COMPARISON WITH OTHER COATING METHODS

The calendering process imparts an even coating within a range of about 0.1–1.5 mm. The upper limit is restricted by the formation of blisters on the coated fabric due to entrapment of air in the rolling bank. For lower thickness, the load requirement on the calender is heavy, moreover, at lower thickness, void formation takes place. An extruder can produce thicker gauges suitably; but, in

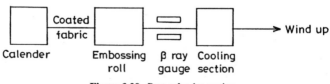

Figure 3.29 Post-calender section.

thinner gauges, there is abrupt variation in thickness [6,13]. The advantage of the calender over an extruder is its (a) high production rate, (b) better product thickness control, and (c) suitability for continuous operation. Compared to fluid coating by spreading, calender coating is a much cleaner process, does not require removal solvents, and uses raw materials that are lower in cost, particularly for vinyl compositions. The cost of calender equipment is, however, higher than either extrusion or spreading [6,14].

3.7.7 ZIMMER COATING

The Zimmer coater (Zimmer Plastics GmbH, Germany) and Bema coater (A. Manrer, Switzerland) are calender-like machines that have been specifically designed for coating fabrics. Thermoplastic polymers in the form of granules, dry powder, or plastic stock are the feedstock for these machines. These machines are less expensive and require lesser manpower and space than a calender.

The Zimmer coater (Figure 3.30) consists of two melt rolls (1) and(2). The rolls are made of diamond-polished, deep-hardened high-grade steel. The gap between the two rolls is adjusted hydraulically. The material is fed at the nip of the rolls, the temperature of which is about 200°C. The coating material melts and adheres to roll (2) which runs at a higher speed and is maintained at a higher temperature. After heating by passage through one or more preheater rolls and IR heaters, the textile substrate is fed between the nip of roll (2) and the backup roll (3). After coating, the hot laminate is either smoothened or embossed by

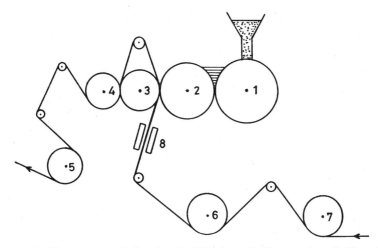

Figure 3.30 Zimmer coater: (1, 2) melt rolls, (3) backup roll, (4) embossing roll, (5) cooling roll, (6) substrate preheat roll, (7) fabric roll, and (8) IR heaters. (Adapted with permission from D. Zickler. *Journal of Coated Fabrics*, Vol. 8, Oct. 1978. ©Technomic Publishing Co., Inc. [14].)

an embossing roll (4). The coated material is then cooled by cooling drums and wound [16].

3.8 HOT-MELT COATING

3.8.1 EXTRUSION COATING

In this process, an extruder converts solid thermoplastic polymers into a melt at the appropriate temperature required for coating. This melt is extruded through a flat die vertically downward into a nip of the coating rolls (Figure 3.31).

The two rolls at the nip are a chromium-plated chill roll and a soft, high-temperature-resistant elastomer-coated backup roll. The chill roll is water cooled. The heat transfer should be adequate to cool the coated fabric so that it can be taken out of the roll smoothly. Means are provided to adjust the position of the die and the nip in three directions. The chill roll may be polished, matt finished, or embossed. Lamination can be accomplished by introducing a second web over the chill roll. The molten resin acts as an adhesive. Extrusion coating is especially suitable for coating polyolefins on different substrates. Because polyolefins can be brought down to low viscosity without risk of decomposition, very high coating rates are achieved, and as such, the process is highly economical. For other polymeric coatings like PVC, PU, and rubber, this process does not yield uniform coating across the width, particularly at thickness below 0.5 mm.

In this method, the coating width can be adjusted by reducing the aperture of the die by insertion of shims. Thus, it is possible to coat different widths for a given die, however, the coating width cannot be changed while coating. Moreover, the process does not permit easy changeover of material. This restricts its use for coating industrial fabrics [16]. A view of an extrusion coating plant is shown in Figure 3.32.

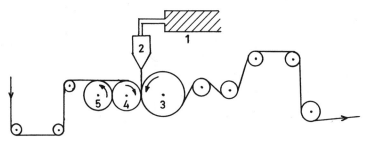

Figure 3.31 Extrusion coating: (1) extruder, (2) die, (3) chill roll, (4) backup roll, and (5) pressure roll. (Adapted with permission from *Encyclopedia of Polymer Science & Engineering*, Vol. 3, 2nd Ed. 1985, ©John Wiley & Sons.)

Figure 3.32 Extrusion coating plant. Courtesy Egan Davis Standard Corp., U.S.A.

3.8.2 DRY POWDER COATING

There are two processes in this category, scatter coating and dot coating [17,18]. These processes are used for coating fusible polymer powder. They are polyethylene, polyamide, polyester, and EVA. The products are used for fusible interlinings, carpet backcoating, especially in the automotive industry for contoured car carpets, and for lamination. The process lends itself to the lamination of two different types of webs, e.g., textiles to foam. Laminates produced by this process retain their flexibility and porosity. Scatter coating is also used for fiber bonding of nonwovens.

In the scatter coating process, polymer powder of 20–200 μm size is spread uniformly onto a moving textile substrate. The web is then passed through a fusion oven and calendered. The method of scattering the powder may be a vibrating screen or a hopper with a rotating brush arrangement, the latter being more accurate. The coating weight is dependent on feed rate and web speed (Figure 3.33).

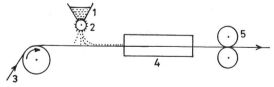

Figure 3.33 Scatter coating: (1) hopper, (2) rotating brush, (3) fabric let off, (4) IR heater, and (5) two-roll calender. (Adapted with permission from *Encyclopedia of Chemical Technology*, Vol. 6, 3rd Ed. 1979, ©John Wiley & Sons.)

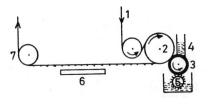

Figure 3.34 Powder dot coating: (1) fabric, (2) oil-heated drum ~200°C, (3) engraved roll, (4) powder feed, (5) cleaning brush, (6) IR heaters, and (7) chilled drum. (Adapted with permission from C. Rossito. *Journal of Coated Fabrics*, Vol. 16, Jan. 1987. ©Technomic Publishing Co., Inc. [18].)

In the powder dot coating process, a heated web having a surface temperature slightly less than the melting point of the polymer is brought in contact with an engraved roller embedded with dry powder. The web is thus coated with a tacky polymer powder in a pattern dependent on the engraving. The engraved roller is kept cool to prevent the polymer from sticking to the roll. A schematic diagram is shown in Figure 3.34 [18].

A new method for printing a hot-melt on a web has been recently described by Welter [19]. In this process, hot molten polymer contained in a trough attached to an engraved roller is picked up by the latter and is pressed into a running web, with the pressure coming from a backup roll. Lamination can also be achieved with another web in the same machine. Various patterns can be printed on the web, including the conventional dot and computer dot printing (Figure 3.35).

Hot-melt coating offers certain advantages over fluid coating. These are as follows [20]:

(1) It does not pollute the environment as no volatiles are emitted.

(2) The rate of production is higher, as it is not dependent on rate of drying/curing.

(3) The plant space requirement is smaller as drying ovens are not required.

(4) The energy consumption is low.

(5) It has better storage stability than fluids.

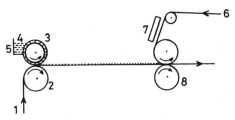

Figure 3.35 Engraved roller melt printing: (1) fabric, (2) backup roll, (3) engraved roll, (4) polymer melt, (5) trough, (6) second web, (7) preheater, and (8) calender. (Adapted with permission from C. Welter. *Journal of Coated Fabrics*, Vol. 24, Jan. 1995. ©Technomic Publishing Co., Inc. [19].)

3.9 REFERENCES

1. G. R. Lomax, *Textiles,* no. 2, 1992, p. 18.
2. Spread coating processes, R. A. Park, in *Plastisols and Organosols,* H. A. Sarvetnick, Ed., Van Nostrand Reinhold, New York, 1972, pp. 143–181.
3. W. R. Hoffman, *Journal of Coated Fabrics,* vol. 23, Oct., 1993, pp. 124–130.
4. Coated fabrics, B. Dutta, in *Rubber Products Manufacturing Technology,* A. K. Bhowmik, M. M. Hall and H. A. Stephens, Eds., Marcel Dekker, New York, 1994.
5. F. A. Woodruff, *Journal of Coated Fabrics,* vol. 21, April, 1992, pp. 240–259.
6. *Encyclopedia of PVC,* L. I. Nass, Ed., vol. 3, Marcel Dekker, New York, 1977.
7. A. Harrera, *Journal of Coated Fabrics,* vol. 20, April, 1991, pp. 289–301.
8. Wire wound rod coating, D. M. MacLeod, in *Encyclopedia of Coating Technology,* D. Satas, Ed., Marcel Dekker, Inc. New York, 1991.
9. R. Greer, *Journal of Coated Fabrics,* vol. 24, April, 1995, pp. 287–297.
10. V. E. Keeley, *Journal of Coated Fabrics,* vol. 20, Jan., 1991, pp. 176–187.
11. *PVC Plastics,* W. V. Titow, Elsevier Applied Science, London, New York, 1990.
12. G. W. Eighmy, *Journal of Coated Fabrics,* vol. 12, April, 1983, pp. 224–236.
13. *Rubber Technology and Manufacture,* C. M. Blow, Ed., Butterworths, London, 1971.
14. J. I. Nutter, *Journal of Coated Fabrics,* vol. 20, April, 1991, pp. 249–265.
15. *Poly Vinyl Chloride,* H. A. Sarvetnick, Van Nostrand Reinhold, New York, 1969.
16. D. Zickler, *Journal of Coated Fabrics,* vol. 8, Oct., 1978 pp. 121–143.
17. M. H. Luchsinger, *International Textile Bulletin,* 1/89, 1989, pp. 5–16.
18. C. Rossitto, *Journal of Coated Fabrics,* vol. 16, Jan., 1987, pp. 190–198.
19. C. Welter, *Journal of Coated Fabrics,* vol. 24, Jan., 1995, pp. 191–202.
20. J. Halbmaier, *Journal of Coated Fabrics,* vol. 21, April, 1992, pp. 201–209.

CHAPTER 4

Physical Properties of Coated Fabrics

4.1 GENERAL CHARACTERISTICS

COATED textiles are flexible composites, consisting of a textile substrate and a polymeric coating. The coating may be on one side or on both sides with the same or a different polymeric coating per side. A typical construction, coated on both sides, is depicted in Figure 4.1.

The physical properties of a coated fabric depend on the properties of the substrate, the coating formulation, the coating technique, and the processing conditions during coating. The factors responsible for different properties of a coated fabric are given in Table 4.1 [1].

Due to the application of longitudinal tension during the coating process, the position of the yarns in the textile substrate is considerably altered in both the warp and weft directions. The warp yarns are aligned more parallel, whereas in the weft there is an increase in the crimp. The minimum coating thickness is thus on the top of the filling yarns.

4.2 TENSILE STRENGTH

The strength of a fabric depends on type of fiber, fineness, twist, and tenacity of yarns and also on the weave and yarn density (set). Theoretically, the tensile strength of a fabric should be the sum of the tensile strength of all the yarns added together. However, there is always a loss of strength due to weaving, and as a result, the theoretical strength is never achieved. The conversion penalty due to weave has been calculated and reported in a recent study on plain weave polyester fabric of varying yarn densities [2]. It has been found that the processing penalty in the warp direction is about 10% and in the weft direction is about 15%. The processing penalty increases with yarn density of the fabric.

The reason for this conversion loss is due to the thread strain during the weaving process, i.e., shedding, warp formation, weft insertion, etc., and due to

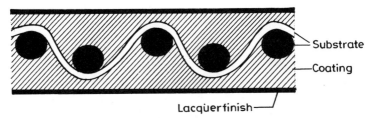

Figure 4.1 A double-sided coated fabric.

the transversal strain at the intersection points. The higher weaving penalty in the weft is due to greater waviness of the weft yarn bending around the stretched warp yarns.

Earlier workers observed that coating increased tensile strength [3–5], however, the reasons for the same were not properly explained. This aspect was thoroughly investigated by Eichert by measuring the strength of coated polyester fabric of different yarn densities [2]. It was observed that the tensile strength of the coated fabric increased from loomstate fabric when calculated on the basis of nominal thread count. Eichert argued that because the difference of break elongation of the yarn and the coating compound is so great, the coating compound cannot contribute to the tensile strength in any way. On further investigation, it

TABLE 4.1. Factors Affecting the Properties of Coated Fabrics.*

Properties	Substrate Construction	Coating Technique/ Processing Conditions	Recipe
1. Tensile strength	•	•	
2. Extension at break	•	•	
3. Dimensional stability	•	•	
4. Burning behavior	•		•
5. Long-time properties	•	•	•
6. Coating adhesion	•	•	•
7. Tear strength	•	•	•
8. Bending resistance	•		•
9. Cold resistance			•
10. Heat resistance			•
11. Chemical resistance	•		•
12. Sea water resistance			•
13. Weather resistance	•		•
14. Abrasion resistance			•
15. Welding properties			•

*Adapted with permission from U. Eichert, *Journal of Coated Fabrics*, vol. 23, April 1994, ©Technomic Publishing Co., Inc. [1].

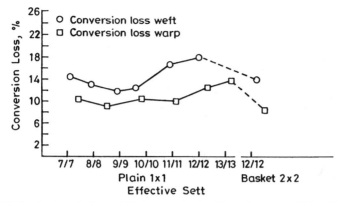

Figure 4.2 Tensile strength of coated fabric conversion loss % vs. effective set (Diolen 174 S 1100 dtex 1210 Z 60). (Adapted with permission from U. Eichert. *Journal of Coated Fabrics*, Vol. 24, July 1994. ©Technomic Publishing Co., Inc. [2].)

was found that there is a shrinkage of fabric in the weft direction due to tension and heat of the coating compound during the coating process. This resulted in an increase in yarn density in the warp direction. A comparison of tensile strength of coated fabric, with that of loomstate fabric, computed from effective thread count showed conversion loss in all fabric densities studied in the warp direction. The conversion loss was found to be more than the weaving penalty. As per Eichert, other reasons for conversion loss due to coating are transversal strain and heating of the yarns in the coating process. The conversion loss is shown in Figure 4.2.

4.3 ELONGATION

The extension of break of coated polyester fabric of different fabric densities has been studied by Eichert (Figure 4.3) [2]. In the uncoated fabric, the elongation only shows a slight increase with thread count both in warp and weft directions. In coated fabric, however, the elongation in the warp direction is much less than that in the loomstate fabric and is similar to the yarn elongation. This is due to the fact that during the coating process, the fabric is subjected to longitudinal tension, stretching the warp threads taut and parallel. Due to the stretching of the warp threads, the looping angle of the weft thread increases, and this phenomenon increases with thread count. This pronounced looping or crimp transfer causes enhanced elongation in the weft direction. Lower elongation of basket weave both in loomstate and coated fabric in weft is due to much lower interlacing.

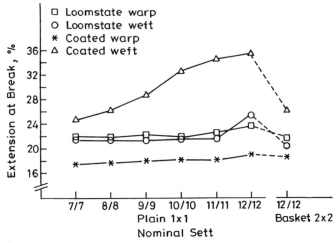

Figure 4.3 Elongation of loomstate vs. coated fabric (Diolen 174 S 1100 dtex 1210 Z 60). (Adapted with permission from U. Eichert. *Journal of Coated Fabrics*, Vol. 24, July 1994. ©Technomic Publishing Co., Inc. [2].)

4.4 ADHESION

The forces of adhesion between the coating compound and the textile substrate are a combination of mechanical and chemical bonding, particularly when bonding agents are added. Mechanical adhesion is predominant in staple fiber yarn and in texturized yarns. Acceptable adhesion values can be achieved between coating and these yarns without the addition of adhesion promoters described earlier. In the case of smooth, high tenacity filament yarns, mechanical adhesion is much lower, and chemical adhesion predominates. Chemical adhesion is obtained by the interaction of the adhesive system with the polar group of the textile substrate and the polymeric coating composition. The effect of mechanical bonding is seen from a study of adhesion in PVC-coated polyester fabric of varying yarn densities (Figure 4.4) [2]. In very loose constructions, the adhesion is very high due to mechanical factors, because of strike-through of the coating composition through the interstices of the fabric. The adhesion decreases with fabric density and becomes more or less level as coating penetration decreases. Thus, even though adhesion on the yarn is at a low level, high adhesion due to mechanical factors can be achieved in scrim fabric.

Chemical adhesion is mainly brought about by treatment of the textile or incorporation of bonding agents in the coating material. The mechanism of the action of the bonding agents has already been described in Chapter 1. Certain coating materials such as polyurethanes and chloroprene also contain reactive groups that promote adhesion. The adhesive system for rubber has already been

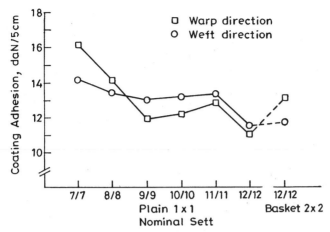

Figure 4.4 Effect of yarn density on adhesion (Diolen 174 S 1100 dtex 1210 Z 60). (Adapted with permission from U. Eichert. *Journal of Coated Fabrics*, Vol. 24, July 1994. ©Technomic Publishing Co., Inc. [2].)

described. The adhesive systems used in trade for coating different polymeric formulations can be classified as follows:

(1) One-component system—polyfunctional isocyanates, e.g., Desmodur R (Bayer), Vulcabond VP (ICI), etc., suitable for rubbers, PVC, and polyurethanes

(2) Two-component system—containing polyols and diisocyanates, e.g., Desmodur N, Plastolein, etc., suitable for PVC and polyurethanes

(3) Three-component system—RFL systems, mainly for rubbers

In vinyl coating, the bonding system is added to the plastisol in a range of about 4–6%. These bonding agents may be one- or two-component systems. The incorporation of the additives increases the viscosity of the PVC paste. Moreover, the plastisol temperature affects the pot life of the bonding agent.

Vulcabond VP [6] is an important bonding agent, for vinyl and textile substrates made from synthetic fibers. Chemically, it is a trimer of toluene diisocyanate dispersed in dibutyl phthalate. The reactive -NCO reacts with the hydroxyl and amido groups of polyester and nylon, respectively, and with the active hydrogen atom of the polyvinyl chloride chain, thus promoting adhesion. Optimum dose level of the additive is ~4%. Higher bonding agent enhances adhesion but causes loss of tear strength of the coated fabric.

The factors affecting adhesion can be summarized as follows [7]

- type of fiber
- fiber surface: moisture, finish, etc.
- construction of fabric

- polymer for coating and its recipe
- bonding agents
- method of coating and coating conditions

Haddad and Black [8,9] have studied the effect of type of yarn, yarn construction, and fabric structure on adhesion. The fabric surface to be tested was pressed against a standard vinyl film. The peel tests were performed as per ASTM 751-79. Effect of weave pattern on adhesion is given in Figure 4.5.

It is seen that adhesion is the same on both sides of a 2 × 2 twill being a balanced fabric. In 1 × 3 twill, however, adhesion is higher on the filling side. This is because the picks were unsized and made of core-spun yarns. The effect of yarns on adhesion was studied by making a series of 1 × 3 twill construction with textured air-entangled polyester as warp and by using different types of filling yarns, viz., multifilament, core-spun, and spun yarns. The highest adhesion was obtained in air-textured yarns, yarns with cotton as the sheath, and open-end spun yarns in the three categories, respectively (the results are given in Table 4.2).

In the case of textured polyester yarns, a relationship between shrinkage (dependent on crimp and bulk) and adhesion has also been found. Clearly, adhesion is dependent on weave and nature of yarn. An open-weave structure and higher yarn surface exposure promote adhesion due to increased mechanical anchoring of the coating compound.

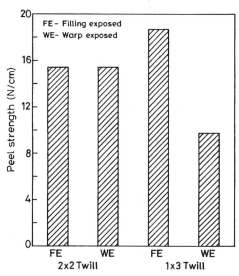

Figure 4.5 Effect of weave pattern and side of laminating on peel strength. Construction: warp—2 × 150/34 textured air-entangled polyester 22.4 ends/cm.; fill—12/1 core spun, sheath 100% pima cotton; core—150/34 stretch polyester, 16.8 picks/cm. (Adapted with permission from Haddad and Black. *Journal of Coated Fabrics*, Vol. 14, April 1985. ©Technomic Publishing Co., Inc. [8].)

TABLE 4.2. Effect of Different Filling Yarns on the Peel Strength on Vinyl Laminated Fabric.* Vinyl Film Applied to the Side Where Filling Is Exposed.

Type of Filling Yarn	Peel Strength N/cm ASTM D 751–79
A. Multifilament polyester yarn	
1. 2 × 150/34 Friction-textured air-entangled polyester	7.9
2. 2 × 150/34 Friction-textured, plied 2.5 TPI S	9.9
3. 2 × 150/34 Air-textured Taslan	13.3
4. 2 × 150/68 Air-textured Taslan	14.9
B. Core-spun yarns	
1. 17.5/1 Sheath polyester, core 150/34 textured polyester	13.8
2. 17.5/1 Sheath AvrillII; core 150/34 textured polyester	17.0
3. 15/1 Sheath 100% pima cotton, core 150/34 textured polyester	18.5
4. 12/1 Sheath 100% pima cotton, core 150/34 textured polyester	18.6
C. Spun yarns	
1. Ring-spun 12/1 3.5TM S high-tenacity polyester	17.0
2. Ring-spun 12/1 2.75TM S high-tenacity polyeser	14.3
3. Ring-spun 12/1 3.5TM S crimped polyester	18.0
4. Ring-spun 12/1 2.75TM S crimped polyester	16.8
5. Ring-spun 16/1 3.5TM Z regular tenacity polyester	17.0
6. Ring-spun 18/1 3.75TM S regular tenacity polyester	13.9
7. Open-end spun 18/1 3.75TM S regular tenacity polyester	23.2

* Adapted with permission from Haddad and Black, *Journal of Coated Fabrics*, vol. 14, April 1985, ©Technomic Publishing Co., Inc. [8].
Construction: 1 × 3 twill, 16.8 picks/cm. Warp 2 × 150/34 textured air entangled polyester 22.4 ends/cm.

Dartman and Shishoo [10] have carried out adhesion of PVC on various polyester and polyamide knitted and woven fabrics. Studies on the effect of moisture were done by carrying out the coating in an environmental chamber with humidity control. Moisture content in the substrate greatly affects adhesion; environmental moisture, on the other hand, has little effect on adhesion (Figure 4.6). They also noted that knitted fabric with a lower cover factor promotes adhesion.

4.5 TEAR RESISTANCE

Resistance to propagation of tear of a coated fabric is of great importance where these fabrics are under tension, e.g., in covers, shelters, and architectural purposes. If cut or punctured, tear can propagate rapidly under stress, damaging the material and leading to its failure. Factors controlling tear strength are as follows:

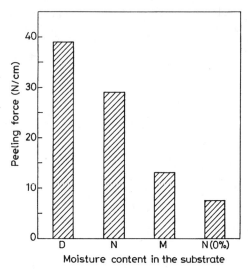

Figure 4.6 Effect of adhesion on the moisture content of substrate. PVC-coated woven polyester, 5% bonding agents in the coating; D = fabric dried at 105°C, N = fabric conditioned at 20°C and 65% RH, M = wet fabric, N (0%) = same fabric as N but with no bonding agent in coating. (Adapted from Dartman and Shishoo. *Journal of Coated Fabrics*, Vol. 22, April 1993. ©Technomic Publishing Co., Inc. [10].)

- construction of the fabric: weave, yarn fineness, and yarn density (Tear strength is related to yarn strength)
- coating material: formulation and bonding system
- adhesion and penetration of coating material on the textile substrate

The effect of various constructional parameters of the fabric on tear strength has been reviewed [11]. A study of three woven constructions, viz., matt 3/3, matt 2/2, and plain-weave fabric, shows that tear strength decreases in the order described due to a lower number of threads at the intersection. Staple fiber yarn has a lower tear compared to filament yarn. Tear strength decreases with weft density of the fabric, but it increases with warp density up to a maximum value and then decreases. The factors that affect tear strength of uncoated fabric also apply to coated fabrics.

Abbott et al. [4] have carried out a detailed study of the tear strength of various woven cotton fabrics coated with PVC plastisols by the knife on blanket method. In the uncoated state, the tear strength of different weaves, with the same cover factor, was found to decrease in the order of basket, twill, and plain weave, due to the reduced deformability of the structures. Coating resulted in loss of tear strength in all cases, but the loss of strength varied with the type of weave. For basket weave, the loss was ~70%, in twill ~60%, and in plain weave ~25%. Yet, the tear strength of the coated fabric was highest for basket, intermediate for twill, and least for plain weave, because of their respective tear strength in

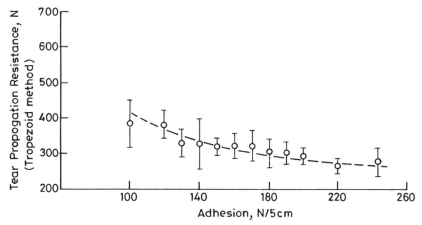

Figure 4.7 Relationship between tear propagation resistance and adhesion (vinyl-coated polyester 1000 d Plain weave 9 ends/cm). (Adapted with permission from Mewes. *Journal of Coated Fabrics*, Vol. 19, Oct. 1989. ©Technomic Publishing Co., Inc. [7].)

the uncoated state. It was also observed that twills and basket weaves made from plied yarns had significantly lower tear strength than those made from single yarns. In all cases, the decrease in tear strength is related to reduced deformability due to the coating.

One of the most important factors controlling the tear strength in a coated fabric is the adhesion of the coating material on the substrate. Mewes [7] has reported a relationship between tear strength and adhesion (Figure 4.7). A linear relationship has been reported by other authors [11,12].

Effect of fabric density on tear strength on PVC-coated fabric has been studied by Eichert [2]. The tear strength in the warp direction decreases with yarn count, but in the weft direction, no general trend is seen (Figure 4.8). In a trapezoid tear test, an increase in ends and picks increases tear strength. This is due to the difference in tear geometry in the two tests. In leg tear, the thread system exposed to testing is not restrained as in the case of the trapezoid test.

The loss in tear strength on coating is much more severe in the weft direction than in the warp. This is explained by the process of spread coating. During spread coating, the coating knife runs parallel to the weft, its dragging action opens the filament/fibers, leading to greater penetration in the weft. Warp yarns, on the other hand, have much lower crimp due to tension applied during coating, moreover, because the knife runs at right angles to the warp, they are not opened, as such, penetration is lower in the warp [3,13]. Abbott et al. [14] have studied the mechanical properties, particularly tearing strength, of different types of cotton fabrics by coating with PVC plastisols of different hardnesses and viscosities; polyvinyl butyral and polyurethane. It was seen that the tear strength is affected primarily by the coating that penetrates the fabric pores, and not on the nature of coating. No direct relation between viscosity and extent of

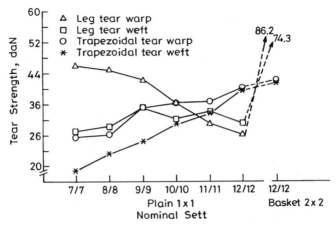

Figure 4.8 Tear strength vs. yarn density in coated fabric. (Adapted from U. Eichert. *Journal of Coated Fabrics*, Vol. 24, July 1994. ©Technomic Publishing Co., Inc. [2].)

penetration could be found. Higher tear strength is observed for coating that is more deformable. In many cases, the penetrated coating was found to be porous and presumably deformable. The use of soft coating alone, or as a base coat with a hard top coat, showed higher tear strength. A novel way of obtaining high tear strength is to fill the fabric interstices with a water-soluble polymer, such as carboxymethyl cellulose prior to plastisol coating, followed by leaching the water-soluble polymer. Thus, high tear strength could be obtained by any type of coating if the shape of the coating-fabric interface and deformability of the penetrated coating are properly controlled.

The studies were continued by the authors to investigate the tear strength and penetration in the coated fabric using different coating techniques, viz., knife on blanket, floating knife, reverse roll, and transfer coating [15]. The tear strength obtained by knife on blanket and floating knife were found to be similar, even though the degree of penetration in floating knife was higher. In transfer-coated fabric, considerable penetration was noted using a binder coat without gelling, but little penetration occurred when the same was pregelled. The reason for higher tear strength, in spite of deep penetration, has been explained as due to the porous, deformable nature of penetrated coating. The factors that determine the tear strength of the coated fabric are the shape of the coating applicator, the nature of the fabric surface, and the rheology of the coating.

4.6 WEATHERING BEHAVIOR

Weathering or degradation of material to outdoor exposure is a complex combination of various components [16].

- solar radiation (In sunlight, ultraviolet radiation is the most important cause of degradation because of its higher energy)
- temperature
- humidity and precipitation—liquid or solid
- wind
- chemicals and pollutants

A coated fabric is essentially a polymeric material. Weathering degrades a polymeric coating by the following process:

a. Volatalization of plasticizer and solvents

b. Rupture of the main macromolecular chain

c. Splitting of the side groups in various ways

d. Formation of new groups and reactions among them

e. Regional orientation—formation of crystalline regions

Studies on the weathering behavior of fibers and fabrics show a rapid degradation in most natural and synthetic fibers resulting in loss of strength [16]. Attempts to correlate outdoor exposure to accelerated weathering tests have not been very successful. A typical result is given in Figure 4.9. The degradation of fabric can be prevented by the application of coating. The thicker the coating, the longer the protection.

Vinyl-coated fabrics are extensively used for covers, tents, shelters, and architectural use. They are, therefore, exposed to weathering. A study of their weathering behavior assumes great significance. The weather resistance of coated fabric is dependent on the coating composition as well as on the nature

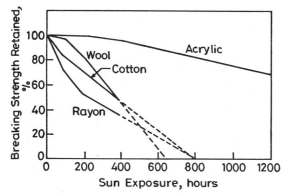

Figure 4.9 Sunlight exposure of wool, cellulosic, and acrylic fibers. (Adapted with permission from "Fibres" by J. H. Ross in *Environmental Effects on Polymeric Materials*, Rosato and Schwartz, Eds. ©John Wiley & Sons, 1968 [16].)

of the textile substrate. Strength and elongation are chiefly contributed by the textile, and the coating protects it from UV radiation during weathering and subsequent loss of strength.

Krummheuer [17] has investigated the weathering of vinyl-coated polyester at different locations of varying sunshine hours and different climatic conditions. The coating composition was a white plasticized PVC with fungicide and UV stabilizer added. The exposure was done at Miami (U.S.), Dormeletto (Italy), Ebnit (Austria), and Wuppertal (Germany). From the results of exposure after five years (Table 4.3), it was found that there was a significant loss of tensile strength and tear strength in thinly coated samples. The loss was maximum at those sites where complete coating was lost, i.e., Dormeletto and Wuppertal. In thickly coated samples, there was loss of coating, but there was no major loss of strength. The rate of loss of strength and tear strength is related to sunshine hours. The rate of deterioration is very rapid after a substantial loss of coating thickness. The author has suggested that for long-term durability, a minimum coating of about 150 μ is required. The outdoor test does not correlate with accelerated weathering by Xenotest because pollutants play an important role. The effect of a UV absorber is significant for short-time exposure, but for long-time exposure, their effect is marginal.

Studies were also carried out by Eichert [1] with similar fabric at the same locations for a period of ten years. The findings are similar to those of Krummheuer. The loss of tensile and tear strength was found to be more severe in the weft direction than in the warp. This is because the fabric warp is located at the center and is better protected than the weft threads.

4.7 MICROBIOLOGICAL DEGRADATION

It has been observed that PVC-coated fabric, used for awnings and marquees, shows discoloration on prolonged use. This is more prominent in white or light-colored fabrics. Such discoloration is due to microbiological attack on the material. In PVC-coated polyester fabric, the plasticizer of PVC is susceptible to microbiological attack, as many of the plasticizers are known nutrients for microbes. Eichert [18] has studied the microbiological susceptibility of white plasticized PVC with and without fungicide. The organisms for study were *Aspergillus niger, Penicillum funiculosum, Paecilomyces varioti, Trichoderma longibrachiatum,* and *Chaetomium globosum.* Profuse growth was observed in coated fabric without fungicide but no growth was seen in uncoated fabric and coated fabric with fungicide. The mechanical properties (tensile strength and tear strength) of the infested fabric, however, did not show any change despite discoloration. The author suggests a longer test to confirm no loss of mechanical damage due to microbiological attack.

TABLE 4.3. Weathering Test for Five Years of PVC Coated Polyester (Plain-Weave Polyester Fabric 1000 d 9/9 Set Coated with White Plasticized PVC).*

| | | | | Original Coating Thickness of Samples and (Add on g/m²) −0.5% UV Absorber | | | | | |
| | 20 μ (540) | | | 50 μ (600) | | | 230 μ (900) | | |
Locations	Res. Thickness	Res. Strength	Res. Tear	Res. Thickness	Res. Strength	Res. Tear	Res. Thickness	Res. Strength	Res. Tear
Wuppertal	0	37%	22%	0	37%	28%	213	113%	87%
Ebnit	22	53%	31%	36	72%	52%	212	87%	94%
Dormeletto	0	21%	14%	0	21%	17%	138	109%	101%
Miami	13	35%	21%	31	51%	36%	204	97%	139%

* Adapted with permission from Krummheuer, *Journal of Coated Fabrics*, vol. 13, 1983, ©Technomic Publishing Co., Inc. [17].

4.8 YELLOWING

For critical applications, yellowing of coated fabrics may be undesirable. In PVC-coated fabric extensively used for architectural application, the yellowing can be due to degradation of the polyvinyl chloride due to heat and light. If properly stabilized, the yellowing of PVC-coated fabric may only be due to the adhesives used in the tie coat formulations. As discussed earlier, these bonding agents are one-component or two-component systems based on polyisocyanates that can be aliphatic or aromatic. Aromatic isocyanates are known to yellow on exposure to light. This aspect has been studied by Eichert [19]. White PVC-coated fabrics with different tie coat formulations were prepared, and their whiteness and light transmission properties were measured after accelerated weathering. The adhesive system affects the yellowing of the fabric, but the aliphatic isocyanates do not show significant nonyellowing properties over many aromatic systems.

4.9 REFERENCES

1. U. Eichert, *Journal of Coated Fabrics,* vol. 23, April, 1994, pp. 311–327.
2. U. Eichert, *Journal of Coated Fabrics,* vol. 24, July, 1994, pp. 20–39.
3. C. L. Wilkinson, *Journal of Coated Fabrics,* vol. 26, July, 1996, pp. 45–63.
4. N. J. Abbott, T. E. Lannefeld, L. Barish and R. J. Brysson, *Journal Coated Fibrous Material,* vol. 1, July, 1971, pp. 4–16.
5. E. H. Mattinson, *Journal Textile Institute,* vol. 51, 1960, p. 690.
6. A. P. Harrera, R. A. Metcalfe and S. G. Patrick, *Journal of Coated Fabrics,* vol. 23, April, 1994, pp. 260–273.
7. H. Mewes, *Journal of Coated Fabrics,* vol. 19, Oct., 1989, pp. 112–128.
8. R. H. Haddad and J. D. Black, *Journal of Coated Fabrics,* vol. 14, April, 1985, pp. 272–281.
9. R. H. Haddad and J. D. Black, *Journal of Coated Fabrics,* vol. 16, Oct., 1986, pp. 123–138.
10. T. Dartman and R. Shishoo, *Journal of Coated Fabrics,* vol. 22, April, 1993, pp. 317–325.
11. *Polymer Modified Textile Materials,* J. Wypych, Wiley Interscience, New York, 1988.
12. Coated fabric, B. Dutta, in *Rubber Products Manufacturing Technology,* A. K. Bhowmik, M. M. Hall and H. A. Benarey, Eds., Marcel Dekker, New York, 1994.
13. V. K. Hewinson, *Journal Textile Institute,* vol. 53, 1962, p. 766.
14. N. J. Abbott, T. E. Lannefeld, L. Barish and R. J. Brysson, *Journal Coated Fibrous Material,* vol. 1, Oct., 1971, pp. 64–84.
15. N. J. Abbott, T. E. Lannefeld, L. Barish and R. J. Brysson., *Journal Coated Fibrous Material,* vol. 1, Jan., 1972, pp. 130–149.

16. Fibres, J. H. Ross, in *Environmental Effects on Polymeric Materials,* vol. 2, D. V. Rosato and R. T. Schwarz, Eds., Interscience, New York, 1968.

17. W. Krummheuer, *Journal of Coated Fabrics,* vol. 13, Oct., 1983, pp. 108–119.

18. U. Eichert, *Journal of Coated Fabrics,* vol. 24, July, 1994, pp. 77–86.

19. U. Eichert, *Journal of Coated Fabrics,* vol. 24, Oct., 1994, pp. 107–116.

CHAPTER 5

Rheology of Coating

$\mathbf{T}$HE science of deformation and flow of matter is termed rheology. It is concerned with the response of a material to an applied stress. In coating, we are concerned mainly with the flow of liquids, solutions, dispersions, and melts. An understanding of the flow property of the coating material is required to control coating thickness, penetration, adhesion, and coating defects.

5.1 RHEOLOGICAL BEHAVIOR OF FLUIDS [1–3]

In order to understand the concept of viscosity, which is the resistance of a liquid to flow, let us consider a situation in which a liquid is confined between two parallel plates—AB and CD. The bottom plate AB is stationary, while the upper plate CD moves (Figure 5.1).

Let the plates be separated by a distance x and the shear force F act tangentially on the top movable plate CD of area A, in a direction, so that plate CD slides sideways with a velocity v as shown in Figure 5.1. The top layer of liquid then moves with the greatest velocity, and the intermediate layers move with intermediate velocities. The velocity gradient dv/dx through the layer is constant, where dv is the incremental change in velocity corresponding to a thickness, dx, of the liquid layer. This term is known as shear rate and is given by

$$\gamma = dv/dx \text{ (shear rate)}$$

The shearing force acting over the unit area is known as the shear stress.

$$\tau = F/A \text{ (shear stress)}$$

Viscosity is defined as the ratio of shear stress to shear rate, i.e.,

$$\eta = \tau/\gamma \text{ (viscosity)}$$

121

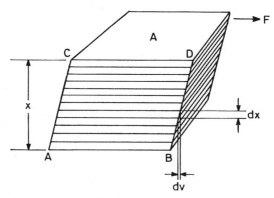

Figure 5.1 Flow of liquid under shear: A = area, F = force (dynes).

The unit of shear stress is dynes/cm^2, shear rate sec^{-1}, and of viscosity dynes $-$ sec/cm^2 or poise.

Liquids where shear stress is directly proportional to shear rate are known as Newtonian. For a liquid, a plot of shear rate vs. shear stress is a straight line passing through the origin. In other words, the viscosity η for a Newtonian liquid is constant, remaining unchanged with rate of shear (Figure 5.2).

Liquids with viscosity that is not constant but varies as a function of shear rate are known as non-Newtonian liquids. The viscosity value obtained for a non-Newtonian liquid at a particular shear rate is known as the apparent viscosity. For non-Newtonian liquids, because viscosity changes with shear rate, only a viscosity profile is capable of expressing the varying viscosity behavior. The various modes of non-Newtonian behavior are given below.

5.1.1 BINGHAM BODY BEHAVIOR

In this type of flow, a certain minimum stress is necessary before flow begins. This is known as the yield value. Once the yield value is reached, the behavior is Newtonian. Mathematically, it can be expressed as $\tau = \tau_0 + \eta\gamma$, where τ_0 is the yield stress. Examples of material of this type are ketchup and mayonnaise.

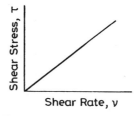

Figure 5.2 A Newtonian liquid.

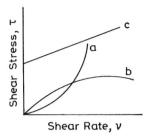

Figure 5.3 Flow behavior of non-Newtonian liquids: (a) dilatant, (b) pseudoplastic, and (c) Bingham body.

5.1.2 DILATANCY AND PSEUDOPLASTICITY

In a dilatant liquid, the apparent viscosity increases with shear rate, i.e., shear stress increases with shear rate. A pseudoplastic liquid, on the other hand, shows shear thinning, i.e., a decrease of apparent viscosity/shear stress, with shear rate. In a dilatant fluid, the dispersed molecules or particles are compressed and piled up, with application of shear, creating resistance to flow. The molecules/particles in a pseudoplastic fluid arrange themselves in a favorable pattern for flow on application of shearing force. The different flow patterns of non-Newtonian fluids and change in apparent viscosity with shear are given in Figures 5.3 and 5.4.

Different mathematical relationships have been put forward to describe non-Newtonian flow behavior. An equation, commonly referred to as the Power law equation, has been accepted to be of general relevance and applicabilty. This equation takes the following form:

$$\tau = K(\gamma)^n$$

where K and n are constants. In logarithmic form, this law takes the form of

$$\log \tau = \log K + n \log \gamma$$

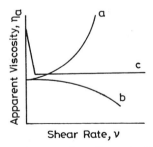

Figure 5.4 Apparent viscosity vs. shear rate of non-Newtonian liquids: (a) dilatant, (b) pseudoplastic, and (c) Bingham body.

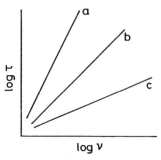

Figure 5.5 Power law plot of different types of fluids: (a) dilatant $n > 1$, (b) Newtonian $n = 1$, and (c) pseudoplastic $n < 1$.

Thus, a plot of log τ vs. log γ will yield a straight line. If $n = 1$, the liquid is Newtonian. Dilatant and pseudoplastic materials have $n > 1$ and $n < 1$, respectively (Figure 5.5).

5.1.3 THIXOTROPY AND RHEOPEXY

There are a number of fluids whose flow properties, such as apparent viscosity, change with time at a constant rate of shear. In some cases, the change is reversible or at least after cessation of shear, the viscosity returns to the original value over time. A thixotropic material may be considered a special case of pseudoplasticity, where the apparent viscosity also drops with time at a constant rate of shear (Figure 5.6). As the shear force is reduced, the viscosity increases but at a lesser rate, forming a hysteresis loop. The area of the hysteresis loop is a measure of the thixotropy of the coating (Figure 5.7). Thixotropic behavior may be visualized as isothermal gel-sol-gel transformation of a reversible colloidal gel. Under constant shear, the material undergoes a progressive breakdown of structure with better flowability over time.

A common example of thixotropic behavior is the drop of viscosity of paint with stirring. Thixotropic behavior is advantageous in a coating system because lowering of the viscosity during coating facilitates application, whereas, higher viscosity at a lower shear rate prevents sagging and dripping.

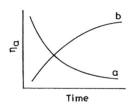

Figure 5.6 Time-dependent flow: (a) thixotropic and (b) rheopectic.

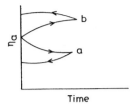

Figure 5.7 Hysteresis curves of of fluids: (a) thixotropic and (b) rheopectic.

Rheopexy is the exact opposite of thixotropy in that under steady shear rate, viscosity increases. This phenomenon is observed in some dilatant systems. Such flow is not of much interest in coating.

5.2 RHEOLOGY OF PLASTISOLS [3–5]

Rheological properties of the coating material are of primary importance for successful coating. The flow properties of coating compositions are greatly influenced by the shear applied during the coating process, as in calender, roller, blade coating, etc. For use of plastisols (pastes) in spread coating, it is useful to know the viscosity at high shear and low shear rates. High shear is encountered at the coating head. A high viscosity at the coating head may cause uneven deposition and may even bend the coating blade. Viscosity at low shear rate and knowledge of yield value is also important. A high yield value prevents strike through in an open-weave fabric, while a low yield value aids leveling of the paste after coating [6]. The rheology of plastisol is the most complex and merits discussion.

Very dilute dispersions containing more than 50% plasticizer show Newtonian behavior. Paste formulations, however, have higher polymer loading. As such, pastes show non-Newtonian behavior. Depending on the formulation, they may be pseudoplastic, dilatant, or thixotropic. The flow behavior normally varies with the shear rate. A paste may show pseudoplastic flow at low shear rates, dilatancy at moderate shear, and again pseudoplastic at still higher shear rates. It may also show dilatancy at low shear rate but pseudoplastic at moderate shear. Pseudoplastic behavior is due to breakdown of a structure in the paste with shear, while dilatancy occurs due to peculiar particle size distribution that does not favor close packing, thus resistance to flow.

5.2.1 APPARENT VISCOSITY OF PLASTISOLS

The viscosity of a simple suspension is given by the well-known Einstein's equation.

$$\eta_s = \eta_o \left(1 + 2.5\phi\right) \tag{1}$$

where η_s = viscosity of suspension, η_o = viscosity of the suspending fluid, and ϕ = the volume fraction of the suspended particles. η_s/η_o is known as the relative viscosity η_r. This equation is true for dilute solutions. At volume fractions >0.025, the η_r becomes much lower than actually observed. Moreover, this equation is applicable when the suspending particles are monodisperse in nature, and there is no interaction between the particles and the medium.

The equation cannot be applied to PVC pastes because of the following reasons:

a. The volume fraction of the suspended particles is quite high >0.2.

b. The suspended particles are not monodisperse.

c. Although paste polymers are resistant to solvation by the plasticizer, slow swelling and dissolution of the polymer particle still occur. Solvation of the polymer increases the viscosity of the plasticizer medium, and swelling increases the volume fraction of the polymer.

Johnston and Brower have developed an equation of apparent viscosity for PVC pastes that is applicable for volume fraction of about 0.2 and for several resin and plasticizer systems.

$$\log_{10} \eta_r = (1.33 - 0.84\ \phi/\phi_c)(\phi/\phi_c - \phi) \qquad (2)$$

ϕ_c is known as critical volume fraction and is defined as the volume fraction of the polymer particle at a stage when it has absorbed plasticizer to the maximum limit, as in a fairly advanced stage of gelation.

The flow property of the paste and its stability are greatly influenced by the formulation of the paste. The important factors are particle size, particle size distribution of the resin, the nature of the plasticizer, and the amount of plasticizer. Additives also affect paste rheology. The effects of various factors are given below.

5.2.2 POLYMER SIZE AND SIZE DISTRIBUTION

As has been described earlier, paste resins are emulsion grade, obtained by spray drying, of particle size ranging from 0.1–3 μm. These particles are known as primary particles. However, during spray drying, aggregates of primary particles are formed that can be much larger in size, 40–50 μm. The aggregates are fragile and break down to primary particles by shearing during paste formation. The ease of breakdown depends on the fragility of the aggregate. The deagglomeration process also affects the paste rheology.

The paste viscosity depends on the size of the primary particles and the size and proportion of the aggregate or secondary particles. The resins may be categorized into the following three types:

(1) High-viscosity resins: the primary particles have size <0.5 μm and are monodispersed in nature. The secondary particles do not significantly affect viscosity.

(2) Medium-viscosity resins: primary particles are polydisperse, having size range of 0.8–1.5 μm. Secondary particles have some effect on viscosity, showing dilatancy at high shear rate.

(3) Low-viscosity resins: they have broader particle size distribution. The secondary particles are larger in proportion and size.

The blending resins have much larger particle size (80–140 μm) than the paste resins. These resins improve the packing of the nearly spherical emulsion particles, reducing the interstitial space and lowering the surface area. Thus, for a given plasticizer content, more of it is freely available, lowering the viscosity.

Apart from particle size, there are other factors of the resin that influence the viscosity of the paste. The surfactant used in the manufacture of the resin is retained in the dried resin. The nature and quantity of the surfactant present in the resin is important as it may reduce or increase the paste viscosity depending on its solubility characteristics, with the plasticizer. Moreover, higher temperature and duration of drying may produce an over-dried resin with reduced plasticizer absorption. The K value of the polymer does not have much influence on the paste viscosity.

5.2.3 PLASTICIZER AND ADDITIVES

The viscosity of the plasticizer and its solvating power affect the viscosity of the paste. For freshly made paste, the viscosity of the paste varies linearly with the viscosity of the plasticizer. However, as the paste ages, this correlation is lost due to the overriding effect of solvation. The higher the solvation (δ closer to PVC $\sim$ 9.7), the greater the viscosity of the paste. Polar plasticizers, such as DBP and TCP, yield highly viscous pastes, with poor viscosity stability, showing dilatancy. For coating purposes, high viscosity and dilatancy are not required. Less polar types such as DOP and DIDP form medium-viscosity pastes with thixotropic properties. As such, they are useful for spread coating. Because of their low δ value, dialkyl esters of adipate and sebacates such as DIDA form low viscosity pastes. An increase of temperature increases aging and viscosity.

The influence of stabilizers, fillers, thickening agents, etc., on paste rheology has already been described (Chapter 1).

5.2.4 VISCOSITY CHANGE DURING FUSION

As the paste is heated, there is initially a lowering of paste viscosity due to a drop in viscosity of the plasticizer with temperature. This is dependent on the nature of the plasticizer (AB in Figure 5.8).

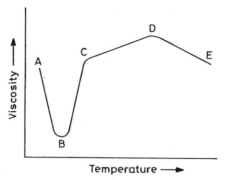

Figure 5.8 Change in viscosity during fusion. (Adapted with permission from *PVC Plastics* by W. V. Titow. ©Kluwer Academic Publishers, Netherlands, 1990 [3].)

A sharp rise of viscosity then occurs mainly due to adsorption of plasticizer by the polymer and due to solution of polymer in the plasticizer-gelation region (BC). The temperature at which the sharp rise occurs is known as the gelation temperature. As the temperature increases, viscosity increases slowly, showing a maxima (D) at the fusion point. A slight drop in viscosity thereafter, is due to the melting of the microcrystalline structure of the polymer. A solvent immersion test is useful for determining complete fusion in a coated fabric. The fusion temperature depends on the nature of the resin and the plasticizer. A paste of a highly solvating plasticizer and fine particle resin of lower K value have lower fusion temperature and require lower fusion time.

5.3 HYDRODYNAMIC ANALYSIS OF COATING

The process of fluid coating is essentially a fluid in motion, and usual unit operation parameters can be applied. An overall macroscopic force balance is obtained by the application of the principle of the conservation of momentum to an elemental volume in the fluid. The force acting on the volume is given by rate of change of the momentum of the fluid surrounding it at any instant, i.e., flux of momentum summed over the entire control surface and the rate of change of momentum within the volume.

The net force F_x, acting in say x direction on the fluid element moving with the velocity of the fluid is a sum of (1) force due to the weight of the volume element (body force) F_{xB} and (2) force due to the stresses acting on it along x direction, F_{xS} [Equation (3)].

$$F_x = F_{xB} + F_{xS} \tag{3}$$

For an element of a differential mass $\rho dx\, dy\, dz$, Equation (3) becomes

$$\rho dx\, dy\, dz(du_x/dt) = \rho dx\, dy\, dz\, g\cos\beta$$
$$+ (\partial\tau_{xx}/\partial_x + \partial\tau_{yx}/\partial_y + \partial\tau_{zx}/\partial_z)dx\, dy\, dz \quad (4)$$

which, on rearranging, reduces to Equation (5),

$$\rho du_x/dt = g\cos\beta + (\partial\tau_{xx}/\partial_x + \partial\tau_{yx}/\partial_y + \partial\tau_{zx}/\partial_z) \quad (5)$$

where

u_x = velocity of the fluid element in x direction

ρ = density of the fluid element

g = acceleration due to gravity

β = angle the fluid element makes with the x axis

t = time

$\tau_{xx}, \tau_{yx}, \tau_{zx}$ = components of stress acting in x direction

By substituting the values of stress and rearranging, we get the Navier Stokes equation [Equation (6)]. This is the equation of motion of the elemental volume in x direction and is used for the analysis of the hydrodynamics of coating [7].

$$u_x\partial u_x/\partial x + u_y\partial u_x/\partial y + u_z\partial u_x/\partial z + \partial u_x/\partial t$$
$$= g\cos\beta - 1/\rho\; \partial p/\partial x + \eta/\rho(\partial^2 u_x/\partial x^2 + \partial^2 u_x/\partial y^2 + \partial^2 u_x/\partial z^2)$$
$$+ 1/3\, \eta/\rho\; \partial/\partial x(\partial u_x/\partial x + \partial u_y/\partial y + \partial u_z/\partial z) \quad (6)$$

where u_x, u_y, u_z are velocities of the fluid element in x, y, and z directions; p = pressure generated due to the movement of fluid element in x direction; and η = viscosity of the fluid.

In a blade coating, the substrate to be coated moves under tension, below a blade. The coating fluid is either poured manually or pumped at the blade nip. The gap between the blade and substrate controls the coating thickness. A study of the hydrodynamics of blade coating has been done by Hwang using the Navier Stokes equation [8]. In a simple analysis, the coated film thickness is a function of five variables only. These are the gap between blade and the web, web speed, viscosity, density, and surface tension of the coating fluid. The motion of the liquid is considered steady and unidimensional in nature, and the liquid is incompressible. For motion along x axis, with y axis perpendicular to it, Equation (6) reduces to

$$\partial p/\partial x = \eta\, \partial^2 u_x/\partial y^2 + \rho g \quad (7)$$

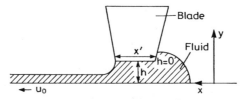

Figure 5.9 Blade coating: h is gap between blade and web; x' is blade width; at the blade $h = 0$, $u = 0$; at the web $y = h$ and $u = u_0$.

The boundary conditions taken are: at the blade, the fluid is motionless, i.e., $u = 0$, the fluid velocity at the web is the same as that of the web velocity, $u = u_0$ at $y = h$. The gap between the blade and the web is h and is small compared to the blade width x' (Figure 5.9).

The coater is more like a channel, and a parallel plane flow model is used. For a Newtonian fluid, by integrating equation [Equation (7)] twice and applying the above boundary conditions, the velocity is obtained as

$$u = u_0 y/h + 1/2\eta \, (dp/dx - \rho g)(y^2 - hy) \tag{8}$$

The total quantity of fluid that passes through the gap per unit length, per unit time, Q is obtained by integrating the above velocity [Equation (8)] between the limits $y = 0$, $u = 0$ and $y = h$, $u = u_0$.

$$Q = \int_0^h u \, dy = u_0 h/2 - (1/12\,\eta)\,(dp/dx - \rho g)h^3 \tag{9}$$

The coating thickness, W, can be obtained by dividing Q by web velocity u_0.

$$W = h/2 - (1/12\eta u_0)(dp/dx - \rho g)h^3 \tag{10}$$

The author has related the pressure gradient to the surface tension of the coating fluid considering the force balance across a parabolic meniscus at the exit of the blade gap and obtained the relation as

$$dp/dx = -\sigma/2h^2 \tag{11}$$

where σ is the surface tension of the fluid.

Thus, the coating thickness obtained in terms of the five paramaters mentioned above is obtained by combining Equations (10) and (11).

$$W = h/2 + (1/12\eta)(\sigma/2h^2 + \rho g)h^3/u_0 \tag{12}$$

The conclusion of the analysis is that for blade coating, the coating thickness is half the coating gap, together with a small term that relates directly to surface tension and gap and inversely to the viscosity and web speed. Hwang has found a good correlation between the coating thickness obtained from experiments and the thickness calculated from Equation (12).

Middleman has carried out analysis of roll, blade, and dip coating for Newtonian and non-Newtonian fluids [9]. These models have, however, not considered the nature of the textile web, which has an important bearing on the coating thickness.

5.4 REFERENCES

1. *Flow Properties of Polymer Melts*, J. A. Brydson, George Goodwin, Ltd. 1981.
2. *Paint Flow and Pigment Dispersion*, T. C. Patton, Wiley Interscience, New York, 1964.
3. *PVC Plastics*, W. V. Titow, Elsevier Applied Science, U.K., 1990.
4. *Manufacture and Processing of PVC*, R. H. Burges, Ed., Applied Science Publishers, U.K., 1982.
5. *The Technology of Plasticizers*, J. Kern Sears and J. R. Derby, Wiley Interscience, New York, 1982.
6. Spread coating processes, R. A. Park, in *Plastisols and Organosols*, H. A. Sarvetnick, Van Nostrand Reinhold, New York, 1972.
7. *Momentum Heat and Mass Transfer*, G. O. Bennett and J. E. Myers, Tata, McGraw Hill, 1962.
8. S. S. Hwang, *Chemical Engineering Science*, vol. 14, 1979, pp. 181–189.
9. *Fundamentals of Polymer Processing*, S. Middleman, McGraw Hill, New York, 1977.

CHAPTER 6

Fabrics for Foul Weather Protection

6.1 CLOTHING COMFORT

A person feels comfortable in a particular climatic condition if his energy production and energy exchange with the environment are evenly balanced, so that heating or cooling of the body is within tolerable limits. A core body temperature of approximately 37°C is required by an individual for his well-being. The body maintains this temperature at different work rates and climatic conditions by changing blood flow and evaporating perspiration from the skin. Because the body has a limited ability to cope with the climate, clothing is consciously selected and adjusted to secure comfort and protection in an adverse environment. There are two aspects of clothing comfort, skin sensorial, i.e., mechanical contact with textile surface, and thermophysiological. The thermophysiological aspect considers the heat balance of the microclimate created between the skin, air, and clothing, with the external climate and the metabolic heat generated. The routes of heat loss from the body are conduction, radiation, and evaporation. The environmental factors responsible for this heat flow are (1) temperature difference, (2) air movement, (3) relative humidity, and (4) radiant heat from the sun or other sources of thermal radiation. Clothing interacts with the environment by [1–3]

- thermal resistance—insulation
- resistance to evaporation
- resistance to wind penetration
- structural features, such as, thickness of clothing, clothing weight, clothing surface area, etc.

The total heat transfer through the clothing of the body with the environment, considering the thermal and evaporation resistance of the clothing, has been

133

given by Woodcock:

$$H = \frac{T_s - T_a}{I} + \frac{P_s - P_a}{E} \tag{1}$$

where H = total heat transfer, $T_s - T_a$ = temperature difference between skin and ambient, $p_s - p_a$ = water vapor pressure difference between skin and ambient, I = insulation of the clothing, and E = evaporation resistance of the clothing.

The thermal and water vapor resistance are additive, and clothing assemblies can be evaluated by adding the individual values and the intervening air layers between them. A person wearing light clothing engaged in light activity in a temperate environment loses about 75% of his metabolic heat by transfer of dry heat. The rest is lost by evaporaton of water from the skin and lungs. As the activity level rises, perspiration production increases, and the proportion of heat loss by evaporation increases. Perspiring is the main thermoregulatory process at a high level of work rates. If the clothing is impermeable, evaporative cooling cannot occur, and in a hot and humid climate, heat exhaustion may occur. Insensible perspiration is converted into liquid perspiration below the dew point. In cold climates, wet, perspiration-soaked clothing loses much of its insulation value, leading to hypothermia. Moreover, wetting of garments by perspiration gives a clinging appearance and is a burden on motion. Such a situation is acute in garments with fabrics having compact coated PVC, PU, or rubbers. Typical perspiration values for various activities are given in Table 6.1.

In extremes of climatic conditions, particularly cold weather, apart from proper thermal insulation, the clothing should be windproof so that cold wind does not enter into the space between the skin and the clothing, dissipating the warm air in the vicinity of the skin. Protection against rain, sleet, or snow is also required, as penetration of moisture in the skin gives a clammy feeling, and its evaporation takes away body heat of the wearer, creating conditions of freezing. In sum, cold weather clothing, besides insulating, should ideally have three main features, it should be water vapor permeable, windproof and waterproof. Two types of fabrics are in use for foul weather clothing. They are

TABLE 6.1. Typical Perspiration Levels for Various Activities [4].*

Activity Level	Heat Produced (watts)	Perspiration Rate (ml/h)
Sleeping (cool dry)	60–80	15–30
Walking 5 km/h	280–350	200–500
Hard physical work (hot humid)	580–1045	400–1000
Max. sweat rate (tolerated for short time)	810–1160	1600

*Adapted with permission from G. R. Lomax Textiles, no. 4 1991, © Shirley Institute U.K. [4]. Conversion factor 1.16 W/m²h = 1 kcal/m²h.

impermeable coated fabrics and the breathable fabrics. An impermeable fabric is both wind- and waterproof but not water vapor permeable. A breathable fabric, on the other hand, meets all of the features of foul weather clothing and is water vapor permeable.

6.2 IMPERMEABLE COATING

These fabrics function by blocking the pores of the textile material by a compact polymeric coating that forms a physical barrier to wind and water. Even though these fabrics are not breathable, they are comparatively inexpensive, and are widely used for rainwear and foul weather clothing.

Various materials are available for rainwear offering different levels of protection. Generally, single- or double-textured rubberized fabrics are used. A single-textured fabric consists of a coating material of natural or synthetic rubber coated on one side of a base fabric of cotton, viscose, or nylon fabric. A typical fabric has a weight of 250 g/m^2, with a proofing content of 140 g/m^2. A double-textured fabric has a rubber coating in between two layers of cotton or viscose fabric. Such a fabric is heavy (400–575 g/m^2) but gives excellent protection against rain [5]. By providing proper ventilation in rainwear, it is possible to transfer condensed sweat outside, minimizing discomfort to the wearer.

Clothing for protection against extreme cold consists of two parts: an inner insulation material and an outer fabric layer to preserve the insulation from wind or rain. The outer layers are usually PU-, PVC-, or neoprene-coated fabrics. PVC- and neoprene-coated fabrics have some limitations in use in extreme cold conditions. PVC-coated fabrics have poor low temperature properties and are affected by solvents. Neoprene-coated fabrics, on the other hand, are rather heavy. Polyurethane-coated nylon is the fabric of choice because of its light weight, thin coating, and excellent low temperature flexibility [6,7]. Some popular PU-coated nylon fabrics have weights ranging from 100–250 g/m^2 with a PU coating between 10–30 g/m^2. These fabrics are also used by the military for various items of cold weather clothing, such as jackets, trousers, caps, and gaiters. Figure 6.1 shows cold weather clothing for the services made of PU-coated nylon fabric.

6.3 BREATHABLE FABRICS

The main features of a breathable fabric are depicted in Figure 6.2.

Extensive research is being done worldwide to develop fabrics that provide comfort to the wearer, while offering protection against foul weather. Although poromerics for shoe uppers were developed in the 1960s, the search for lightweight breathable fabric for apparel got a fillip from the development

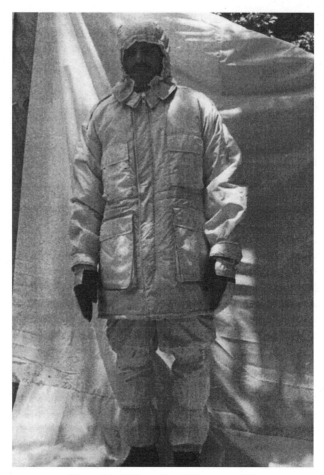

Figure 6.1 Cold weather clothing ensemble.

of the versatile GORE-TEX® laminates in 1976. Numerous brand products have been developed and patents filed ever since. In the ensuing sections, an overview of the technology of these fabrics is discussed.

6.3.1 USES AND REQUIREMENTS

These fabrics find extensive use in sports and leisure wear. Army personnel on outdoor duty are exposed to foul weather for days or weeks, especially when on patrol duty. Breathable fabrics have great application for protective clothing for the services. Another emerging field for breathable fabric is protective apparel for healthcare workers, against body fluids and bacterial and viral infections. ASTM has adopted two new specifications to evaluate the barrier effectiveness of such clothing. They are ASTM F 1670-97 and F 1671-97a [8].

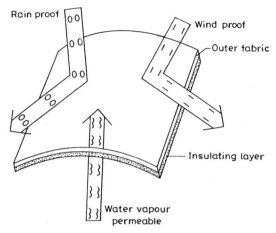

Figure 6.2 Main features of a breathable fabric.

In quantitative terms, a breathable fabric should have the following attributes [4,9,10]:

(1) Water vapor permeability—min. 5000 $g/m^2/24$ h

(2) Waterproofness—min. 130 cm. hydrostatic pressure

(3) Windproofness—less than 1.5 $ml/cm^2/sec$ at 1 mbar; measured by air permeability

Other properties required are as follows:

(1) Durability: tear, tensile, and peel strength; flex and abrasion resistance

(2) Launderability

(3) Tape sealability

6.3.2 DESIGNING A BREATHABLE FABRIC

A discusssion of the mechanism of water vapor permeability and water repellency is useful in understanding the principle of designing breathable fabric. Water vapor transport through a fabric/clothing system may occur due to diffusion (driven by vapor concentration gradients) and convection (driven by pressure difference). A discussion of the mechanism of diffusion of water vapor through fabric and membranes is given below.

6.3.2.1 Diffusion of Water Vapor through a Fabric

This occurs by the following ways [1,11].

(1) Interyarn space: the diffusion of water vapor occurs through these spaces due to the water vapor pressure gradient across the two sides of the fabric by molecular diffusion.

(2) Interfiber space: the contribution of diffusion through a fiber bundle is much less than the void between the yarns. However, liquid water can permeate the fabric by the wicking action of capillaries of the fiber bundle and subsequent evaporation at the outer surface.

(3) Intrafiber diffusion: in this process, water vapor is absorbed by the fiber and desorbed at the outer surface. The process is dependent on the nature of the fiber, i.e., hydrophilic or hydrophobic, and is related to moisture regain. It is obvious that diffusion of air and water vapor follow different mechanisms.

A detailed study of diffusion of water vapor through fabric was carried out by Whelan et al. [12]. To arrive at a theoretical model, studies were done using perforated metal plates of different thickness and hole diameter. It was found that water vapor resistance is directly related to the thickness of the plate and inversely to the percent pore area. For plates having constant perforated area, the resistance to water vapor diffusion increases linearly with the diameter of the perforation. An empirical formula [Equation (2)] was derived from the experimental data.

$$R = \frac{T}{\beta} + 0.71d\left(\frac{1}{\beta} - \frac{1}{\sqrt{\beta}}\right) \tag{2}$$

where R = water vapor resistance, d = diameter of holes, and β = ratio of void area to total area.

Studies of permeation with fabrics revealed that water vapor resistance is related to the thickness of the fabric, provided fiber volumes are similar. In the case of fabrics, because the air path is irregular, the resistance is related to the percent fiber volume (V_f), rather than percent pore areas found in case of metal plate. An empirical relation derived [Equation (3)] fits the experimental data.

$$R = \frac{100}{100 - V_f}(0.9 + 0.034V_f)T + 0.5 \tag{3}$$

The results indicate that fabric thickness and the air space in the fabric are important parameters for movement of water vapor. The diffusion of water vapor through the pores of a polymer foam structure is related to the thickness and percent pore volume and follows the empirical equation of the metal plate as given in Equation (2) [13].

6.3.2.2 Diffusion through Membranes [14]

Diffusion of gases and vapor through a nonporous membrane occurs through a different route. The gas is initially dissolved in the exposed polymer surface. The concentration built up on the surface is directly proportional to gas pressure, i.e., Henry's law. The gas then migrates to the opposite surface under concentration gradient. The migration of the penetrant can be visualized as a sequence of steps during which the molecule passes over a potential barrier separating one position from the next. A successful jump requires that a passage of sufficient size be available, and this is dependent on the thermal motion of the polymer chains. The diffusivity is temperature dependent, which follows an Arrhenius-type expression. For large molecules, the size of the penetrant determines the hole size required. Diffusion rates are higher for good solvents of the polymers, rather than for permanent gases, as they diffuse by plasticizing the polymer. The steady state flux J is given by the expression as

$$J = DS(p_1 - p_2)/l \tag{4}$$

where $D =$ diffusivity, $S =$ solubility coefficient/Henry's law constant, $l =$ thickness of the membrane, p_1 and p_2 are partial pressure of the diffusing gas at two sides of the membrane.

It is common practice to describe the diffusional character of the membrane at equilibrium in terms of a quantity known as permeability. The permeability coefficient is given by $P = DS$. The structure of the polymer has great influence on the permeability. The factors that increase the segmental mobility, enhance the diffusion rate. Thus, permeation is higher at temperatures higher than the glass transition temperature T_g. Increase of structural symmetry and cohesive energy of the polymer decreases the permeation rate. The crystalline domains in the polymer are inaccessible to penetrants and are an impermeable barrier for the diffusion process. The presence of cross-links reduces the segmental mobility, reducing the diffusion rate. The permeation of water vapor varies from polymer to polymer, depending on the presence of certain polar groups (amino, hydroxyl, carboxyl) in the polymer chain that can interact with water molecules, forming reversible, hydrogen bonds. These groups act as stepping stones for transport of water vapor through the polymer. Water vapor initially absorbed acts as a plasticizer, increasing the intermolecular holes, through which nonhydrogen-bonded water molecules can also pass [11].

6.3.2.3 Water Vapor Transport through Textiles

The water vapor diffusion through textiles is determined by diffusion tests such as ASTM-E-96 and ISO-11092. Such tests can lead to erroneous results for high air permeability textile materials, because a very small pressure gradient can produce large convective flows through the porous structure, far outweighing diffusive transport. Therefore, to characterize the potential of a given material to transport water vapor through its structure, it is necessary to carry out both diffusion and air permeability tests.

In recent years, extensive work has been done by Gibson on the water vapor transport properties of textiles [15–18]. He has developed a fully automated test method known as "dynamic moisture permeation cell," that enables determination of diffusion and convective properties from the same test, of diverse types of materials, such as air-impermeable laminates, very high air-permeable knitted fabrics, woven fabrics, and polymer foams.

In the test method, the sample is placed in a cell. Nitrogen streams consisting of a mixture of dry nitrogen and water-saturated nitrogen are passed over the top and bottom surfaces of the sample. The relative humidities of these streams are varied by controlling the proportion of saturated and dry components. By knowing the temperature and water vapor concentration of nitrogen flows entering and leaving the cell, the flux of water vapor diffusing through the test sample is measured. It is also possible to vary the temperature of the cell and pressure drop across the sample. With this test method, Gibson has been able to measure the following parameters of various materials:

(1) Combined convection and diffusion: these studies are done by varying the pressure drop, across the sample. With a specified pressure drop, transport takes place both by diffusion and by convection. If there is no pressure drop across the sample, the transport is only by diffusion. For convective transport of water vapor, Gibson has used Darcy's Law [Equation (5)] for calculation of permeability.

$$v = (-k_D \Delta p)/\mu \Delta x \qquad (5)$$

where v = apparent gas flow, k_D = permeability constant, Δp = pressure difference across sample, and Δx = thickness. The results on various types of fabrics clearly show that for high air permeability textiles, convective water transport dominates.

(2) Humidity-dependent air permeability: this parameter is important for porous hygroscopic materials that often exhibit humidity-dependent air permeability due to the swelling of fiber as it takes up water from the environment.

(3) Concentration-dependent diffusion: vapor transport across nonporous hygroscopic polymer membrane is often highly dependent on the amount of water present in the polymer. Studies on several such membranes confirm the same.

(4) Temperature-dependent diffusion studies: for hydrophilic films and membranes, water vapor transport is significantly affected by temperature. The water vapor transport is lower at lower temperatures. This effect is important for the ability of cold weather clothing to dissipate water vapor during active wear.

(5) Transient sorption and desorption: The method can be used to conduct testing of materials under nonsteady state conditions, such as change in relative humidity, temperature, or pressure difference across the sample. In these transient situations, the variable properties of the material become very important, along with factors like sorption rate at which fiber takes up and releases water.

6.3.2.4 Repellency [19]

A liquid spreads on a solid surface when its surface energy (γ_{SA}) is higher than that of the liquid (γ_{LA}). The spreading process lowers the free energy. The spreading coefficient is given by

$$S = \gamma_{SA} - (\gamma_{LA} + \gamma_{SL}) \tag{6}$$

where γ_{SA}, γ_{LA} are the surface tension of the solid and the liquid in contact with air, and γ_{SL} is the interfacial tension. If the spreading coefficient is positive, spreading occurs, and the contact angle is $<90°$. For low energy solids, the spreading coefficient is negative, the liquid is repelled, and the contact angle is $>90°$. Critical surface tension of a solid surface is equal to the surface tension of a liquid that exhibits zero contact angle on the solid. It is a measure of the repellent property of the solid surface. A repellent functions by lowering the critical surface tension of the solid. For instance, a fluorochemical emulsion lowers the critical surface tension of a nylon fabric to <10 dyne /cm, as such, it can repel water and hydrocarbon oils having surface tension 72 dynes/cm and ~ 20 dynes/cm respectively.

A textile surface is not smooth, liquid can migrate into its pores, even though the contact angle on the surface is $> 0°$. The transport of a liquid through a capillary is given by

$$\Delta P = 2\gamma_{LA} \cos /r \tag{7}$$

where ΔP is the pressure required to force the liquid into the capillary, and r is the pore radius.

Penetration of water into the capillaries of the textile can be prevented by reducing the size of the pores and increasing the contact angle through a water repellent treatment. Water repellent finishes are of various types, viz., pyridinium compounds, wax and wax emulsions, silicones, fluorochemicals, etc. Several commercial finishes are based on molecules containing polar and nonpolar moieties. The polar ends of the molecule attach to the textile, whereas the nonpolar part sticking outwards repels the water. Silicones are available in solution or aqueous emulsion and are commonly a blend of polymethyl hydrogen siloxanes and polydimethyl siloxanes. A variety of fluorochemicals are used as repellents, and several brand products exist. Copolymer of fluoro alkyl acrylates and methacrylates are primarily used for textiles. Fluorochemicals have an added advantage in that they are oil repellent as well.

As mentioned earlier, a compact coated fabric is wind- and waterproof but impermeable to water vapor. The contradictory requirements of water vapor permeability and waterproofness are achieved, in a breathable fabric, by specifically designing the coated fabric to have interconnecting pores of 0.2 to 10 μm. These pores permit much smaller water molecules ~ 0.004 μm to permeate from inside of the clothing and at the same time prevent ingress of much larger water droplets $\sim$100 μm (for a fine drizzle) to the body. Because the size of water molecules are of the same order as those of the constituents of air, these microporous coatings/films are permeable to air. However, their permeability is so low that they impart windproofness to the fabric. Another concept that has emerged recently are hydrophilic coatings. These are compact coatings that transport water vapor by permeation through polymeric membranes as discussed earlier. Breathable fabric can be engineered from closely woven uncoated textiles, known as ventile fabrics in common parlance.

6.3.3 TYPES OF BREATHABLE FABRICS

Breathable fabrics can be categorized into four main types:

(1) Closely woven fabrics with water repellent treatment

(2) Microporous film laminates and coatings

(3) Hydrophilic film laminates and coatings

(4) A combination of microporous coating with a hydrophilic top coat

6.3.3.1 Closely Woven Fabrics

The earliest of the waterproof, water vapor permeable fabrics were the Ventile fabrics developed by the Shirley Institute U.K. [3,11] during World War II. The development was an outcome of an urgent need to protect the survivors of air crew forced to ditch in the cold North Seas from hypothermia. Immersion suits of Ventile fabric are still in use in the U.K. These fabrics are made from long staple Egyptian cotton, using low twist mercerized yarns, woven in a dense oxford construction. The fabric weight ranges from 170 to 295 g/m² for different uses. The interyarn pores of the fabric in dry state are about 10 μm. The air permeability is low, but the interyarn spaces and hydrophilic nature of the fiber allows adequate water vapor permeability. On wetting the fabric, by rain or immersion in water, the cotton yarn swells, reducing the pore size to 3–4 μm. The swollen fabric in combination with repellent finish, prevents further penetration of water by rain or seawater. The choice of repellent treatment is critical, as it should still allow absorption of water by cotton yarn to swell and constrict the interyarn pores. The waterproofness of these fabrics is low and can stand only moderate hydrostatic pressure.

A new generation of high density fabrics made from microfibers of 0.05 to 1 d polyester, polyamide, viscose, or acrylics has recently emerged as breathable fabric with improved functional properties. The microfibers are obtained by melt spinning of two incompatible polymers into a single fine fiber, known as a bicomponent fiber [20,21]. The cross section of the fiber may be either side by side, core sheath, or matrix fibril type. One of the polymers is then separated by dissolving in a specific solvent, leaving behind microfiber. The yarn is woven

TABLE 6.2. Important Microfibers and Fabrics*

Microfiber	Fabric	Supplier
1. Trevira finesse	Clima guard	Hoechst
2. Tactel micro	Micro spirit	I.C.I.
3. Dyna bright	H_2 OFF	Toray
4. Supplex	—	Dupont

*Adapted with permission from M.Van Roey, *Journal of Coated Fabrics,* vol. 21, July 1999. ©Technomic Publishing Co. Inc. [22].

into various dense fabric constructions like taffeta, twill, or oxford, and given a repellent finish of silicones or fluorochemicals. These fabrics have better water repellency than the cotton ventile fabrics and have very soft handle. Some of the important microfibers and fabrics are listed in Table 6.2 [22].

6.3.3.2 Microporous Coatings and Laminates

These are porous membranes laminated to a fabric or porous coating. The pore size ranges from 0.1 to 50 μm. The most widely used are polyurethanes, poytetrafluoroethylenes, acrylics, and polyamino acids. Among these, polyurethane is the most popular polymer because of toughness, flexibility of the film, and capability of tailor making the property of the film to suit the end use requirement. Various methods of generating microporosity have been reported in the literature, and these are discussed below [3].

6.3.3.2.1 Wet Coagulation Process

Microporous polyurethane coating by direct or transfer process has been discussed in detail under poromerics in synthetic leather (Chapter 7). A film, on the other hand, is obtained by casting, using the transfer coating process on a release paper, which is subsequently adhered to the fabric to make laminates. Some commercial products in this class are Cyclone (Carrington), Entrant (Toray), Keelatex, etc. Incorporation of water-soluble salts in the PU coating solution has also been described (Chapter 7). Microporosity is created by leaching the salts on treatment of the film with water. Products in this class are Porvair, Porelle, Permair, etc. [3,11]. Various improvements in the process have been reported in the literature. Addition of a water-repellent agent and nonionic surfactants in the coating solution, imparts water repellency to the pores, resulting in better waterproofness of the coated fabric. A water repellent treatment of the coated layer further improves the waterproofness [23]. A much higher water vapor permeability and waterproofness have been reported by Furuta et al. [24] by incorporation of about 1% nonporous inorganic filler, e.g., silica (Aerosil) or magnesium oxide of particle size <0.1 μm in the polyurethane resin. The coating obtained on wet coagulation shows ultrafine pores of <1 μm, in addition to a honeycomb skin core structure of 1–20 μm pores. The formation of these micropores has been explained as due to the subtle difference in the rates of coagulation at the resin particle interface. The enhanced water vapor permeability (>6000 g/m^2/24 h) and water pressure resistance (>0.6 kg/cm^2) are attributed to the formation of these additional micropores.

The phase separation method for forming microporous coating has been developed by UCB Chemicals, Belgium, for their product Ucecoat 2000 [11]. In this process, polyurethane is dissolved in a solvent mixture of methyl ethyl ketone, toluene, and water, having 15–20% solids and coated on the fabric. The low boiling solvent evaporates leading to precipitation of polyurethane in the nonsolvent. The nonsolvent is then removed by drying, to yield a microporous coated fabric. This process has an advantage over the wet coagulation process in that immersion and washing baths are not required. The number of pores and their size in microporous coating obtained by the wet coagulation process are $\sim 10^6$ pores/cm^2 and 3 to 40 μm, respectively [9].

A combination of polyamino acid (poly-γ-methyl-L-glutamate) and polyurethane resin in the ratio of 60:40 to 40:60 has been used by Unitika Co., Japan, for production of their Exceltech brand of microporous coated fabric by wet coagulation process [25]. An optimum quantity of surfactant is added for improved water vapor permeability by controlling the porosity. The base fabric used is 70 d taffeta. A scanning electron micrograph shows 10^7 pores/cm^2 in the product. Because PAU is hydrophilic in nature, Exceltech has higher moisture permeability, 8000–12,000 g/m^2/24 h, compared to 4000–6000 g/m^2/24 h for microporous PU coating. The other properties are water entry pressure of 200 cm and air permeability of <0.07 ml/cm^2/sec which are comparable to microporous PU coating.

Microporous polyamide coating by wet coagualation process has also been reported [26]. The process consists of application of an hydrochloric acid solution (5–7.5 N) of polyamide containing 20–30% solids on a textile substrate by knife coating. The coated fabric is immediately dipped into 5–10 N caustic soda solution bath, to coagulate the polyamide. The excess caustic soda solution is then squeezed out from the coated fabric by passage through rolls, washed with water, and excess water is removed. To prevent damage of the fabric by the acid solution, the neutralization process is carried out quickly. The coating is about 20 μm thick, containing two distinctict types of pores. The outer layer consists of small ovoid pores of about 0.02 μm due to instant coagulation, while the inner layer has large elongated pores with average diameter of ~ 1 μm.

6.3.3.2.2 Microporous Polyurethane from Aqueous Dispersions

The process consists of impregnation of a textile web with aqueous dispersions of polyurethane containing solubility enhancing ionizable groups and coagulating the polymer by acid or alkali solution, depending on the charge. Such methods are environment friendly as no solvents are used, but the coated film has poor adhesion and durability. Dahmen et al. [27] have developed a method that gives much better adhesion and durability of the coating. The textile is coated with aqueous dispersions of polyurethane having ionic groups of opposite charge, i.e., both anionic and cationic dispersions. Commercially available dispersions are taken, and their viscosity is adjusted by the addition of nonionic thickeners to render them suitable for knife coating. Anionic dispersion is applied as the first coat on the textile, followed by a second coat of cationic dispersion in wet condition, without any intermediate drying or vice versa. The weight ratios of the two dispersions are so adjusted that the anionic and cationic groups are stoichiometrically equivalent. The coated fabric is then air dried at about 140°C, given a flourocarbon treatment, and calendered lightly. The process has an advantage in that existing solvent-based equipment can be used and the time-consuming rinsing process is avoided. The fabric is waterproof and water vapor permeable (water vapor permeability ~2400 g/m^2/24 h).

6.3.3.2.3 Microporous polytetrafluoroethylene [3,22]

Thin films of 5–15 μm PTFE are extruded through a slit die and biaxially stretched. This results in the formation of microtears of pore size 0.1–1 μm and ~10^9 pores/cm^2 in the film. The film is mechanically weak and therefore laminated to textile fabric by adhesives. The film is hydrophobic in nature, and its water repellency is far superior. GORE-TEX® (W. L. Gore, U.S.) brand product is the most widely used and versatile laminate of this type. The process [28] consists of extruding PTFE paste-dispersion in mineral spirit. The extrudate is dried to form a film of unsintered PTFE. The film is then clamped and stretched. The stretching can be done in one direction (uniaxial stretching) or in two directions, right angles to each other (biaxial stretching). The stretching is done at an elevated temperature, below the melting point of the polymer, at a high rate. While still stretched, the film is heated slightly above the melting point of the polymer and cooled rapidly in stretched condition. The process gives a film containing porous microstructure, with a considerable increase in strength. The microstructure of the uniaxially stretched film consists of nodes elongated at right angles in the direction of the stretch. These nodes are interconnected by fibrils that are oriented parallel to the direction of the stretch. Typically, the size of the nodes vary from 50–400 μm. The fibrils have widths of about 0.1 μm and lengths ranging from 5–500 μm. The development of porosity occurs due to void formation between nodes and fibrils. When the films are biaxially stretched, similar fibril formation occurs in the other direction with the production of cobweb-like or cross-linked configurations with an increase in strength. Porosity also increases as the voids between the nodes and fibrils become more numerous and larger in size. The factors affecting the porosity and strength of the film are as follows:

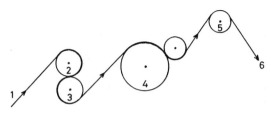

Figure 6.3 Production of stretched PTFE film: (1) unsintered film, (2, 3) heated calender rolls, (4) heated roll for heat treatment, and (5) water-cooled roll.

(1) The polymer should have high crystallinity, preferably >98%.

(2) The temperature and rate of stretching: higher temperature and higher rate of stretch lead to a more homogeneous structure with smaller, closely spaced nodes, interconnected with a greater number of fibrils, increasing the strength of the polymer matrix. Typically, stretching is done between 200–300°C.

(3) The temperature and duration of heat treatment: during heat treatment, which is done above the melting point of the polymer (350–370°C), an increase in amorphous content of the polymer occurs. The amorphous region reinforces the crystalline region, enhancing the strength without substantially altering the microstructure.

Continuous length of the porous film can be obtained by passing the unsintered film through the nip of heated calender rolls moving at different speeds. The film is stretched in the gap of the rolls. The extent and rate of stretch depends on the friction ratio and the gap between rolls. The stretched film coming out of the calender is passed on a heated roll for heat treatment at ~370°C and then over a water-cooled roll for cooling (Figure 6.3). Nitto Elec Ind., Japan, has come out with a similar product known as Microtex.

Figure 6.4 shows an extreme cold weather suit with an outer GORE-TEX® laminate and insulation layer.

6.3.3.2.4 UV/E Beam Polymerized Membranes [29,30]

A rapid method of making microporous films has been developed by Gelman Science, U.S., and patented in 1984 as the Sunbeam process. Monomers and oligomers are cross-linked under a radiation source, UV/E beam and cured in milliseconds. The polymer is based on acrylates. The pores are of the order of 0.2 micron, and the film can be very thin, of the order of a few angstroms. The technology permits production at a rapid rate of about 350 ft/min in 1 m width and is suitable for making films as well as coating. The film is highly repellent and is not wetted by water and most chemicals, only solvents with a S.T. < 18.5 dynes/cm wet the film. The water vapor permeability of these films are ~1100 g/m^2/24 h, and the films can be dry cleaned. These films are

Figure 6.4 GORE-TEX® suit for cold weather.

marketed as "Repel" and can be laminated on nonwovens for making disposable protective clothings for a variety of applications such as clean room garments, medical gowns, chemical splash suits, etc.

6.3.3.2.5 Perforation in Compact Coated Fabric [9,11]

A method has been developed to create micropores of the order of 1 million pores/m^2 by passage of nonporous coated fabric between two electrodes, generating high voltage. The electron beam creates pores through the coating without damaging the fabric. This method gives smooth straight through perforations. The functional properties are therefore poor.

6.3.3.2.6 Extraction of Soluble Component from a Polymer Mixture

A process for making breathable fabric with a microporous polyethylene-tetraflouroethylene (ETFE) layer has been reported by Kafchinski et al. [31]. The process consists of making a dope containing a suspension of ETFE ~0.5–1 μm in an extractable polymeric binder. The dope is cast onto a release paper and dried. The film is next calendered above the flow temperature of ETFE, i.e., about 300°C, and laminated to a fabric substrate using heated calender rolls. The binder is then extracted by a suitable solvent which is a nonsolvent for ETFE and does not damage the fabric. This leaves behind a microporous ETFE layer on the fabric. Various extractable polymeric binders can be used such as polycarbonate, polymethyl methacrylate, etc. However, water-soluble polymer, such as polyethyl oxazoline, cellulose acetate, etc., are preferred because they can be extracted by immersion in water. The fabrics preferred are polybenzimidazole, Kevlar,® Nomex®, etc. The resultant laminated fabric is flame resistant as well as breathable. The microporous film can be bonded to the fabric by adhesive as well.

In another process, fabric is coated with an aqueous-based composition containing a film-forming polymer and a suitable proportion of a water-soluble polymer. The coated film is dipped in an aqueous solution of an enzyme, which degrades the water-soluble polymer. Washing of the degradation product leaves behind a microporous coating. The film-forming compositions are emulsions of acrylics, silicones, polyurethanes, or their mixtures. Water-soluble polymers include starch, carboxymethyl cellulose, sodium alginate, etc. The enzymes used are specific for the water-soluble polymers, e.g., for starch it is amylase, and for cellulose derivatives cellulase is used. The water vapor permeability obtained is of the order of 4500 g/m^2/24 h, which is substantially higher than that obtained on washing the coated film with plain water [32].

6.3.3.3 Hydrophilic Coatings and Films

The mechanism of permeation of water vapor through a nonporous polymer film has been discussed. None of the conventional coatings like PVC, PU, and rubbers have the polar groups required for activating the hydrophilic mechanism for the transport of water. Although a number of hydrophilic polymers, like polyvinyl alcohol and polyethylene oxide are available, they are highly sensitive to water and would either dissolve completely in contact with rain or swell so heavily that the flex and abrasion resistance would be very poor. A proper hydrophilic polymer for coating should, therefore, have adequate swelling to allow transmission of water vapor and at the same time retain suitable film strength. Cellulosic derivatives, with a high percentage of crystallinity, lead to stiff coatings. To achieve the proper functional property of the film, an optimization of hydrophilic-hydrophobic balance is required. The approaches available are the use of polymer blends, incorporation of pendant hydrophilic groups, or the use of segmented copolymers [3]. Out of these possibilities, use of segmented copolymer, incorporating polyethylene oxide into a hydrophobic polymer chain has been found successful. The polyethylene oxide segment has a low binding energy for water, permitting rapid diffusion of water vapor, and it is flexible, so that the end product has soft handle [11,33].

6.3.3.3.1 Hydrophilic Polyurethanes

These are segmented copolymers of polyester or polyether urethanes with polyethylene oxide. The hydrophilicity can be varied either by increasing the overall hydrophilic content or by changing the length of the hydrophilic segments. A process for their synthesis has been described in a patent to Shirley Institute, U.K. [34]. A prepolymer is prepared by reacting polyethylene glycol of molecular weight varying from 1000–2000, with an excess of diisocyanate (preferably 4,4-diisocyanato-dicyclohexyl methane). The prepolymer is chain extended by a low molecular weight diol. Cross-linking can be done using a tri-isocyanate. The polyethylene glycol content is maintained within 25–45 wt.% for optimum properties. An example of this product is Witcoflex [11,33], a series of coatings manufactured by Baxenden Chemicals, U.K. There are two products, viz., Staycool and Super dry. The coating formulations resemble two-component polyurethane systems. The coating can be applied by direct or by transfer coating in MEK/DMF solution. The tie coat contains isocyanate cross-linkers. Bion II film of Toyo also falls in this category [22]. Krishnan [8,35] has reported the synthesis of hydrophilic-polyurethane film based on polyurethane and dimethyl siloxane. Polyalkylene ether glycol (C atoms in the alkylene group at least 3) and polyoxyethylene glycol are reacted with a stoichiometric excess of diisocyanate and chain extended by diamine or diols. In a similar manner, polyalkylene ether glycol, polydimethyl siloxane diol, and polyethylene glycol are reacted with an excess of diisocyanate and chain extended. The two systems are then blended and cross-linked with isocyanates. By controlling the molar ratio of the polyols, a proper combination of hydrophilicity and hydrophobicity is achieved by this process. The product is made by Raffi and Swanson and is marketed as Comfortex. The system is suitable for direct coating as well as for film lamination. The laminate can be used for medical protective clothing. A similar process has been described by Ward et. al. [36]. The soft segment of a block copolymer of polyether urethane consists of a hydrophobic component obtained from either polydimethyl siloxane or polytetramethylene oxide and a hydrophilic part provided by polyethylene oxide. This polymer is mixed with a compatible base polymer, preferably polyether urethane urea. A solution of the two polymers in dimethyl acetamide is suitable for coating or for casting as a film. Incorporation of 0.5% LiBr in the coating solution enhances the water vapor permeability. Another approach of preparing hydrohilic PU is by incorporating polyamino acid in polyester polyurethane. The higher the PAU content, the higher the water vapor permeability, but the lower the elasticity of the film

[37]. Film obtained by transfer coating is laminated on fabric. Excepor U of Mitsubishi is a product of this type [22].

Desai and Athawale [38] prepared a series of PU coating compositions of varying hydrophilicity by incorporating polyethylene glycol of different molecular weight in castor-oil-based polyester polyurethane. These compositions were coated on nylon fabric, and their properties were studied. The water vapor permeability showed an increasing trend with an increase of molecular weight of PEG. In washfastness test, the weight loss of polymer increased with an increase of molecular weight of PEG, indicating that an increase in hydrophilic segments lowers its adhesive strength. In a similar study on cast films by Hayashi et al. [39], it was found that moisture permeability increases linearly with both the concentration of polyethylene glycol and its molecular weight. Setting of the T_g of the polymer at ambient increases its moisture permeability above room temperature but lowers the same at cold temperature. Such shape memory polymers may find applications in foul weather clothing.

6.3.3.3.2 Hydrophilic Polyester

Sympatex membrane devloped by Akzo/Enka Germany [40,41] is a hydrophilic polyester into which polyether groups have been incorporated to impart hydrophilicity to the membrane. Commercial films, 10–25 μm thick, are produced by extrusion process. The membrane is colorless and opaque. It has a water vapor permeability of over 2500 g/m^2/24 h, accompanied by about 5% swelling. It is watertight up to 1000 cm. The film is laminated to the fabric.

Hoeschele and Ostapchenko [42] have synthesized breathable waterproof film of thermoplastic hydrophilic polyetherimide ester elastomers. The elastomeric film is made by the reaction of a diol with a dicarboxylic acid and a polyoxyalkylene imide diacid. The preferred diol and dicarboxylic acid are 1,4-butane diol and terephthalic acid. The polyoxyalkelene imide diacid is obtained by the imidization of trimellitic or pyromellitic acid with polyoxyalkylene diamine. The diamine has molecular weight between 900–4000 and is obtained from polyethylene glycol and polypropylene glycol. The polyoxyalkylene imide diacid can be depicted by the general formula

$$\text{HOOC-R} \underset{\underset{O}{\overset{\|}{C}}}{\overset{\overset{O}{\overset{\|}{C}}}{\diagdown}} \text{N-X-N} \underset{\underset{O}{\overset{\|}{C}}}{\overset{\overset{O}{\overset{\|}{C}}}{\diagup}} \text{R-COOH}$$

Polyoxyalkylene imide diacid: X = polyether chain, R = acid residue

A large stoichiometric excess of diol is reacted with polyoxyalkylene imide diacid and a dicarboxylic acid, in the weight ratio of 0.8 to 3.0. The resulting polyetherimide ester elastomer containing 25–60 wt.% of ethylene oxide has optimum properties. The elastomer is compounded with UV stabilizers and fillers and extruded as film. The film is laminated to a textile substrate. Water vapor permeability of 3500 gm/m²/24 h as per ASTM E 96-66 has been claimed.

Hydrophilic films and coatings have certain advantages over microporous materials [11,33]. In the wet coagulation method, coagulation baths, washing lines, and DMF recovery plants are required. Moreover, precise control over the coating operation is required to generate a consistent, uniform pore structure, preferably below 3 μm for optimum balance of breathability and waterproofness. The entrapment of residues of detergent and sweat into the pores alters the surface property considerably and reduces the water penetration pressure. Hydrophilic coatings, on the other hand, can be applied by conventional solvent coating equipment, being nonporous, they do not lose their properties on cleaning. A hydrophilic film is sometimes applied on microporous films to upgrade the water resistance. Thintech of 3M is a microporous polyolefin coated with a hydrophilic polyurethane. Similarly, UCB Chemicals have also developed a product in which Ucecoat NPU hydrophilic finish is applied to microporous Ucecoat 2000 [22].

6.3.4 USING BREATHABLE FABRIC FOR APPAREL

A breathable coated fabric is used as an ensemble with a liner. The coated surface is always inside, to protect from abrasion. The coating can be on the outer fabric or on the liner. A repellent treatment is generally applied to the outer fabric. Breathable films/membranes are not used as such. They are converted into laminates by bonding them to fabric before use. Lamination is done by powder dot, paste dot, adhesion nets, etc. The resistance of water vapor of these laminates depends on the nature and thickness of the membrane, area of the membrane covered by the adhesive, and nature of the textile component. There are different methods of making laminates. In a garment, the membrane is always the second layer, from outside, placed directly below the outer fabric. The various types of laminates [9,40] are shown in Figure 6.5.

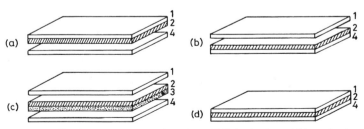

Figure 6.5 Types of laminates: (a) outer fabric laminate, (b) lining laminate, (c) insert laminate, (d) three-layer laminate, (1) outer fabric, (2) breathable film, (3) insert fabric, and (4) lining material. (Adapted with permission from M. Drinkmann. *Journal of Coated Fabrics,* Vol. 21, Jan 1992. ©Technomic Publishing Co., Inc.) [40].

The membrane may be bonded to the outer fabric [Figure 6.5(a)], to the liner [Figure 6.5(b)], on a lightweight knitted material (insert laminates) [Figure 6.5(c)], or may be bonded to both outer and inner fabric into a trilaminate [Figure 6.5(d)]. The type of laminate to be selected depends on the intended application. In the outer fabric laminate, functional aspects are paramount. The liner and insert laminates give softer handle and better fashion appeal.

6.3.5 EVALUATION

The fabric properties of breathable material can be evaluated by B.S. 3546. There are, however, different test methods for measuring water vapor permeability. Because the conditions of tests are different, they give different values. A new specification, B.S. 7209-90, has appeared specially for breathable fabrics [33]. This specification deals with two grades of water resistant, water vapor permeable apparel fabrics. The main requirements of the specification are (1) water vapor permeability index %; (2) resistance to water penetration as received, after cleaning, after abrasion, and after flexing; (3) cold crack temperature; (4) surface wetting (spray rating) as received and after cleaning; and (5) colorfastness to light, washing, dry cleaning, and rubbing. The method of determining the water vapor permeability index has been discussed in the chapter on test methods.

A comparative evaluation of several breathable fabrics available in trade has been done by Keighley [2]. The fabrics have been categorized into three types, viz., Ventile, PTFE film laminates, and PU coating. Some of his important findings are given in Table 6.3. It is seen that cotton Ventiles have low waterproofness but higher permeability. GORE-TEX® laminates have both very high permeability and waterproofness but are expensive.

TABLE 6.3. Comparative Properties of Different Breathable Fabrics.*

| Properties | Fabrics | | |
	Cotton Ventile	PTFE Laminates (GORE-TEX®)	PU Coating
1. Waterproofness hydrostatic head cm	160–200	>2127	630->2127
2. Water vapor permeability British MOD specification upright cup at 35°C. g/m^2/24 h	4100–5150	4850–5550	2500–4650

* Adapted with permission from J. H. Keighley. *Journal of Coated Fabrics,* vol. 15, Oct. 1985. ©Technomic Publishing Co., Inc. [2].

6.3.6 COMPARISON WITH IMPERMEABLES

A comprehensive study has been carried out by several TNO laboratories of The Netherlands on the comfort property of rainwear. Rainwear was made from waterproof fabrics of different technologies, viz., microporosity, continuous impermeable films, and microfiber weaves. In a wear trial, it was shown that breathable garments caused equal thermal strain as impermeable clothing in cold environments but less strain in hot conditions. During hard work, the moisture accumulation in breathable fabrics is lower but still large enough to cause discomfort. However, when worn for a whole day, with continually changing moisture production and climate, a breathable garment dissipates moisture all of the time, whereas in impermeable garments, only accumulation takes place. For extended periods of wear, breathable fabrics give a more comfortable, dry feeling. Ventilation at appropriate places in the garment has been designed as an alternative means to transport water vapor [43].

6.4 REFERENCES

1. *Clothing Comfort and Functions,* L. Fourt and N. R. S Hollies, Marcel Dekker Inc., New York, 1970.

2. J. H. Keighley, *Journal of Coated Fabrics,* vol. 15, Oct., 1985, pp. 89–104.

3. Coated and laminated fabrics, R.A Scott, in *Chemistry of the Textile Industry,* C. M. Carr, Ed., Blackie Acad. Professional, London, 1995.

4. G. R. Lomax, *Textiles,* no. 4, 1991, pp. 12–16.

5. M. A. Taylor, *Textiles,* vol. 11, no. 1, 1982, pp. 24–28.

6. G. R. Lomax, *Journal of Coated Fabrics,* vol. 14, Oct., 1984, pp. 91–99.

7. C. Cooper, *Textiles,* vol. 8, no. 3, 1979, pp. 72–83.

8. K. Krishnan, *Journal of Coated Fabrics,* vol. 25, Oct., 1995, pp. 103–114.

9. W. Mayer, U. Mohr and M. Schuierer, *International Textile Bulletin, Dyeing, Printing, Finishing,* 2/1989, pp. 16–32.

10. S. Krishnan, *Journal of Coated Fabrics,* vol. 22, July, 1992, pp. 71–74.

11. G. R. Lomax, *Journal of Coated Fabrics,* vol. 15, July, 1985, pp. 40–66.

12. M. E. Whelan, L. E. MacHattie, A. C. Goodings and L. H. Turl, *Textile Research Journal,* vol. xxv, no. 3, March, 1955, pp. 197–223.

13. G. F. Fonseca, *Textile Research Journal,* Dec., 1967, pp. 1072–1076.

14. *Polymer Permeability,* J. Comyn, Ed., Elsevier Applied Science Publishers, Ltd., New York, 1985.

15. Water vapor diffusion in textiles, P. Gibson, *Clemson University Coated Fabrics Conference,* Clemson, SC, U.S.A., May 8–9, 1997, pp. 49–92.

16. Water vapor diffusion in coated fabrics, P. Gibson, *Clemson University Coated Fabrics Conference,* Greenville, SC, U.S.A., May 14–15, 1996, pp. 1–28.

17. Water vapor transport properties of textiles, P. W. Gibson, *Clemson University Coated Fabrics Conference,* Clemson, SC, U.S.A., May 11–12, 1999, pp. 1–28.

18. P. Gibson, *Journal of Coated Fabrics,* vol. 28, April, 1999, pp. 300–327.

19. *Waterproofing and Water Repellency,* J. L. Molliet, Elsevier, London, 1963.

20. *Kirk Othmer Encyclopedia of Chemical Technology,* 4th Ed., vol. 10, John Wiley and Sons, New York, 1995, pp. 654–655.

21. J. Hemmerich, J. Fikkert and M. Berg, *Journal of Coated Fabrics,* vol. 22, April, 1993, pp. 268–278.

22. M. Van Roey, *Journal of Coated Fabrics,* vol. 21, July, 1991, p. 20.

23. Y. Naka and K. Kawakami, U.S. Patent 4,560,611, Dec. 24, 1985.

24. T. Furuta, K. Kamemaru and K. Nakagawa, U.S. Patent 5,024,403, April 20, 1993.

25. T. Furuta and S. Yagihara, *Journal of Coated Fabrics,* vol. 20, July, 1990, pp. 11–23.

26. J. L. Guillaume, U.S. Patent 4,537,817, Aug. 27, 1985.

27. K. Dahmen, D. Stockhausen and K. H. Stukenbrock, U.S. Patent 4,774,131, Sept. 27, 1988.

28. R. W. Gore, U.S. Patent 3,953,566, April, 1976.

29. High Performance Textiles, vol. 9, no. 11, 1989, pp. 7–8.

30. E. C. Gregor, G. B. Tanney, E. Schchori and Y. Kenigsberg, *Journal of Coated Fabrics,* vol. 18, July, 1988, pp. 26–37.

31. E. R. Kafchinski, T. S. Chung, W. Timmons and J. Gasman, U.S. Patent 5,358,780, Oct. 25, 1994.

32. T. Tanaka, T. Tanaka and M. Kitamura, U.S. Patent 4,695,484, Sept. 22, 1987.

33. G. R. Lomax, *Journal of Coated Fabrics,* vol. 20, Oct., 1990, pp. 88–107.

34. J. R. Holker, G. R. Lomax and R. Jeffries, U.K. Patent G.B.2,087,909A, 1982.

35. S. Krishnan, U.S. Patent 5,238,732, Aug. 24, 1993.

36. R. S. Ward and J. S. Riffle, U.S. Patent 4,686,137, Aug. 11, 1987.

37. European Pat. Appl. no. 83305387, Sept. 9, 1983, in *Journal of Coated Fabrics,* vol. 14, Jan., 1985, pp. 148–164.

38. V. M. Desai and V. D. Athawale, *Journal of Coated Fabrics,* vol. 25, July, 1995, pp. 39–46.

39. S. Hayashi and N. Ishikawa, *Journal of Coated Fabrics,* vol. 23, July, 1993, pp. 74–83.

40. M. Drinkmann, *Journal of Coated Fabrics,* vol. 21, Jan., 1992, pp. 199–210.

41. *High Performance Textiles,* vol. 7, no. 6, Dec., 1986, pp. 2–3.

42. G. K. Hoeschele, G. J. Ostapchenko, U.S. Patent 4,868,062, Sept. 19, 1989.

43. Institute for Perception, TNO report no. IZ F 1986-26, Oct., 1986.

CHAPTER 7

Nonapparel Coating

7.1 SYNTHETIC LEATHER

A footwear material primarily protects the feet from the environment in which it is worn. Leather is preferred as a footwear material because it has certain desirable properties such as high moisture absorption as well as high air and moisture vapor permeability. The removal of liquid perspiration by evaporation and absorption makes a leather shoe more comfortable to wear. The stretchability of leather makes it suitable for the lasting process, in the manufacture of the shoes. The waterproofness of leather is not adequate. A coating to make it waterproof makes it impermeable. Leather can be given different finishes, viz., patent, grain, and suede.

The search for synthetic leather has been catalyzed due to periodic shortages of leather and for economic considerations. Moreover, natural leather comes in different sizes and thickness, thus it cannot be handled by automatic, computer-controlled production lines. Synthetic substitutes are now increasingly in use for various footwear components like shoe uppers, linings, insole, insole covers, stiffeners, etc. A UNIDO report predicts that in the near future, only 60% of shoe material will be leather, and the rest will be alternate materials. Ideally, a man-made shoe upper material should have a similar appearance, with mechanical and physical properties comparable to natural leather. Synthetic materials for footwear applications are generally coated fabrics. These may be compact coated impermeable fabrics or breathable fabric/poromerics.

7.1.1 COMPACT COATED FABRICS

In the quest for a low cost alternative to leather as a material for shoe, upholstery, and other applications, vinyl-coated fabrics have been considered the most suitable material. The flexibility of the product can be adjusted by varying the plasticizer content, and a leather-like appearance can be given by embossing. The initial PVC fabrics were nonexpanded coatings on closely woven fabrics.

155

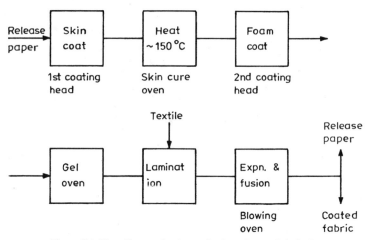

Figure 7.1 Flow diagram for the production of expanded vinyl.

These fabrics are stiff, lack the handle of leather, and are difficult to upholster. Besides, they tend to harden and crack quickly due to migration of the plasticizer. These fabrics typically have a thick skin coat ~190 g/m^2 on plain weave cotton base fabrics of ~150 g/m^2. They find use as upholstery material in less expensive outdoor furniture [1,2].

In the 1950s, expanded vinyl-coated fabrics were developed, which had handle and drape properties very similar to leather. Expanded vinyl fabric consists of knitted fabrics as the base material, an intermediate layer of cellular PVC, and a wear-resistant top coat [1,3]. These fabrics are also known as leather cloth and are produced by transfer and calender coating. In the transfer coating process, a thin coating of plastisol is applied on a release paper or steel belt. This top coat is partially fused at about 150°C. A second coat of plastisol containing 1–5% blowing agent is then applied on the top coat. This layer is again partially fused, laminated to fabric, and then subjected to complete expansion and fusion at about 200–230°C (Figure 7.1). The temperature of the oven depends on the nature of the blowing agent. A glossy top surface can be obtained by applying an acrylic precoat prior to casting the top coat. Embossing can be achieved by using embossed release carrier.

In the clandering process, vinyl sheeting containing a blowing agent is produced in the normal manner. The sheet is then laminated to a wear layer and fabric. The composite layer is then heated in an oven at high temperature (~200°C) to activate the blowing agent to produce the foam layer [4]. A cross section of the various layers of expanded vinyl is shown in Figure 7.2.

Upholstery grade cloth has a thick foam layer ranging from 360–480 g/m^2, a top layer of 180 to 360 g/m^2, embossed to a leather-like grain and a knitted base fabric of ~100 g/m^2. Expanded vinyls are lower in cost and have better

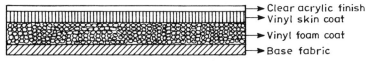

Figure 7.2 Layers of expanded vinyl.

durability than leather. They have, however, a cool feel, undergo cracking due to plasticizer migration, and cannot be dry cleaned [2].

Polyurethane-based fabrics are superior to vinyl-coated fabrics in leather simulation, durability, and low temperature flexibility. These fabrics contain a knitted base fabric, a polyurethane foam middle layer, and a wear resistant top coat similar to the expanded vinyls [3]. The method of their manufacture is also similar. A top coat of PU is cast on a release paper in the transfer process. This is followed by a coating mixture of polyol, isocyanate prepolymer, and a blowing agent. The composite is then foamed, cross-linked, and laminated to a textile base. The foamed imitation leather is then separated from the release paper [5].

Both expanded vinyl and polyurethane-based leather cloth are widely used for upholstery, soft luggages, handbags, shoes, seat covers, and door panels of cars, etc. [1,3]. An important requirement of upholstery fabric is that it should have proper flame retardant additive to reduce the ignitability of the products, smoke generation, and toxicity of the decomposition products [6].

The vinyls have a very low water vapor permeability. The PU-coated fabrics, because of their hydrophilic nature, have some water vapor permeability $\sim$5–18 g/m^2/h, but the same is not adequate enough for comfort properties of shoe materials [7].

7.1.2 POROMERICS

These are second generation synthetic leathers and are so named as they contain porous polymers. Besides aesthetics, they have certain other important characteristics:

a. A microporous structure

b. Air and water vapor permeable

c. Water repellent on the outer decorated side

These properties render poromerics as a shoe upper material having properties in between leather and vinyl-coated fabrics, because all the properties of leather have not been achieved in poromerics as yet.

The most prevalent method of manufacturing [7–9] poromerics is to incorporate a soft porous polymeric mass in the fabric matrix and subsequently coat it with a porous polymeric layer. The porous polymer is obtained by coagulation of a polymer solution in a nonsolvent. The polymer most suited for the purpose is polyurethane. Typically, a solution of polyurethane in dimethylformamide (DMF) is applied on a fabric, by dipping and/or coating, followed by dipping in a large excess of water. The polyurethane coagulates in the nonsolvent due to precipitation and coalescence. On drying, the microporous polymeric material is obtained that imparts both waterproofness and water vapor permeability to the fabric. DMF is used as a solvent for polyurethane because of the following:

- It is a good solvent for PU.
- It has a high boiling point, so it does not readily evaporate from solution.
- It is highly miscible in water. Water can therefore permeate into the DMF solution, bringing about coagulation.

The textile materials used as backing are generally nonwoven fabrics that may be nonreinforced or reinforced with knitted or woven fabric. Nonreinforced nonwoven has poor strength and higher permanent set. Fabrics in woven form are also used in the form of raised pile, however, they suffer from inferior ductility. The fiber properties, particularly its strength, water absorption, and fineness, are important parameters determining the final properties of the end product. Fine fibers, particularly microfibers, give excellent aesthetics comparable to those of leather [10]. The structure of a poromeric with nonwoven backing material resembles that of leather, the microporous top coat and the backing are comparable to the grain and reticular layers of leather. Poromerics can also be obtained by transfer coating on a release paper followed by coagulation and lamination of the microporous film onto a fabric such as in Porvair.

7.1.2.1 Coagulation Process [5,11]

A completely reacted polyurethane solution of about 20% in case of nonwovens or about 10% for a woven substrate is mixed with aqueous pigment, ionic polyurethane (to promote coagulation), and a polyelectrolyte. The solution is deaerated prior to use. The steps involved in the manufacture are varied depending on the manufacturer, but a typical flowchart is given below. The fabric is initially impregnated with the PU solution by coating or dipping, followed by coagulation in a water bath, and excess water removal by squeezing. To the moist PU impregnated substrate, a coating of polyurethane is applied by knife coating. Steam is then passed over the fabric for gelling and to initiate coagulation. The fabric is then dipped into a coagulation bath containing 15–35% DMF in water. After coagulation, the fabric is washed in water bath, followed by washing in suction drums, where DMF is removed and recovered. The wet fabric is next dried in a tenter (Figure 7.3).

The pores of the polymer are of micron level and are interconnecting, leading to permeability. Various factors control microporosity and other properties of the poromerics [7,9].

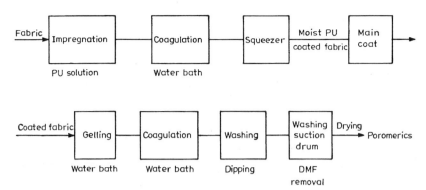

Figure 7.3 Flow diagram of manufacture of poromerics by coagulation.

(1) The type of polyurethane and the additives

(2) Viscosity and solid content of the solution

(3) Type of fabric

(4) Extent of impregnation

(5) Concentration of DMF in bath and temperature of coagulation

(6) Total dip time and extent of squeezing

(7) Washing efficiency

(8) Drying conditions

A variation of the process is the incorporation of water-soluble [5,7] salts like sodium chloride or ammonium sulphate into the PU solution. During co-agulation, the salts leach, forming a controlled pore structure. Microporous polymer can also be obtained by dissolving polyurethane into a volatile solvent (THF) and a high boiling nonsolvent (a hydrocarbon solvent). On evaporation, the volatile solvent leaves behind polyurethane in the nonsolvent, leading to precipitation of the polymer and coagulation. The microporosity depends on the quantity of the nonsolvent. A process of coating fabric with porous PU from PU dispersion has been developed by Stahl-Holland [10]. A fabric is impregnated and coated with aqueous polyurethane dispersion. Precipitation and coagulation is done at an elevated temperature above 93°C, using hot water, steam, or microwave. The end product containing porous polymer has much better leather-like properties. The process is nontoxic and requires a simpler production unit.

7.1.2.2 Poromerics from Prepolymers [5]

The coagulation process described above uses fully reacted polyurethane. Two methods of manufacture of poromerics from prepolymers have been

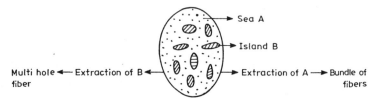

Figure 7.4 Structure of special fiber of Clarino. (Adapted with permission from *Encyclopedia of Chemical Technology*, Vol. 14, 3rd Ed. 1979, © John Wiley & Sons [7].)

reported in the literature. They are polyaddition in solution and dispersion. In the former, prepolymer is dissolved with chain extender in a solvent or solvent mixture in which the end product is insoluble. As the reaction proceeds, the polyurethane formed becomes less soluble with time and precipitates out, occluding the solvent. On evaporation of the solvent, a poromeric is obtained. In the dispersion process, isocyanate prepolymer is dispersed in a solvent, (aromatic hydrocarbons) nonsolvent (water) mixture. To this emulsion, the chain extender is added. Polyaddition occurs during evaporation of the solvent mixture, leaving behind a porous polymeric structure.

The dried fabric containing coagulated PU is then finished by a thin spray coating and embossing in a calender, to impart the desired color and grain for leather-like appearance. Suede-like appearance is obtained by buffing.

A number of brand products of poromerics are available in the market [7,8,12]. They differ in the coating as well as in the textile backing. For instance, Porvair is a microporous film laminated to a fabric substrate, whereas Clarino (Kuraray) contains coagulated polyurethane in a fiber matrix with a top coat of microporous polyurethane. Nonreinforced nonwoven fiber matrix has been used in Clarino. For this purpose, special fiber has been developed by Kuraray that is close in properties to collagen [13]. Two immiscible polymers are melt spun to give a fiber of a cross section shown in Figure 7.4. It has been termed as island and sea fiber where one polymer is dispersed in another polymer matrix. By selective solvent extraction of either the dispersed phase (island) or the matrix (sea), a hollow supple multihole fiber or a bunch of fine denier fiber results. The use of these fibers leads to a product of considerably enhanced suppleness. Contemporary man-made leather such as Sofrina (Kuraray) and Ultrasuede (Toray) employ microfibers of less than 0.3 denier for the nonwoven matrix to obtain excellent properties resembling leather [7].

A cheaper alternative to poromerics [9] is obtained by applying a top coat of conventional coating of polyurethane by the transfer process, on a fabric substrate already impregnated with porous polyurethane by the coagulation process. This method has much better aesthetics, handle and appearance, and seam holding properties than compact PU-coated fabrics but poorer permeability than poromerics. These fabrics are extensively used in Europe for ladies footwear.

7.1.2.3 Structure of Porous Films

The structure of porous films formed by the coagulation process and the effect of various additives have been studied by Chu et al. [14]. They have noticed a dense skin layer on top of a spongy base layer. The base layer consists of finger-like cavities and large cavities. Cellulose acetate membranes used for the desalination process are made by a similar coagulation process using acetone-solvent and water-nonsolvent and have similar porous structure. The reasons of formation of such a structure of the cellulose acetate membranes have been intensively investigated [15–17]. The top skin layer is formed either by the evaporation of the solvent or by rapid precipitation at the outermost surface. As the water diffuses through the skin layer to the precipitation zone, the rate of coagulation slows, and coarser precipitate is formed. Thus, the pore size increases from top to bottom. Similar reason can be attributed to the pore structure of porous polyurethane. The effect of additives studied by Chu et al. [14] are sodium nitrate, water, and defoaming agents. Addition of sodium nitrate and water in the PU-DMF solution accelerates the coagulation process and promotes formation of larger cavities in the base layer. Antifoaming agents such as Span-60 and octadecanol, which are hydrophobic in nature, lower precipitation rate. Dense skin layer is formed with Span-60, and finger-like cavities are absent using octadecanol. It has also been found that temperature of the coagulation bath plays an important role in pore size formation. A cold bath produces a small poromeric structure throughout the entire coating; a warm bath creates a small structure at the surface and a large pore structure underneath. The reason for this is the heat of solution generated when DMF and water are mixed. If the bath is cold, it has the ability to absorb this heat without significant temperature rise. The polymer-rich layer then coagulates quickly, resulting in a small pore structure. If the bath is warm, the water-rich layer penetrates more deeply into the precipitation layer before the polymer cools to the point that it will coagulate. This gives rise to a large pore structure. The other important factor in pore size development is the concentration of DMF in the water/DMF bath. A high DMF concentration yields a large pore structure.

7.1.3 POROUS VINYLS [18]

These materials are used as liners in footwear for absorption of moisture. They are much cheaper than PU-coated fabrics and poromerics. There are various techniques to produce these materials:

(1) Incorporation of soluble material into the plastisol matrix, calendering it into sheet form, leaching the soluble material to produce voids, and then laminating it on to fabric
(2) Foaming (chemical or mechanical)

(3) Sintering—careful heating and pressure sinters solid vinyl particles, generating a solid containing voids

Murphy [18] has described a novel and improved process of manufacturing absorptive vinyl. The process consists of coating a fabric with plastisol containing noncompatible thermoplastic polymer particles. During fusion and gelling, the incompatible thermoplastic softens and shrinks, as it has no adhesion with the vinyl matrix. By controlling the size and content of particles and by applying mechanical stress, it is possible to create interconnecting tunnel-like voids. The absorptive vinyl is laminated to a backing material of natural-synthetic fiber blend and finished. Absorptive vinyls have high moisture absorption and desorption rates and are particularly suited for liner applications using impermeable uppers.

7.1.4 PTFE LAMINATE [19]

GORE-TEX® is a hydrophobic polytetrafluoroethylene film containing micropores. It repels water but permits passage of water vapor. A triple laminate with an outer layer of textured nylon fabric, a middle layer of GORE-TEX® film, and an inner layer of knitted fabric can be used as a shoe upper. Total ingress of water is prevented by sealing rather than stitching the seams for construction of the shoe. These uppers are used for applications where a high degree of water repellency combined with breathability are required.

7.2 ARCHITECTURAL TEXTILES

The use of shelters made of textile material for protection against the elements dates back to mankind's earliest days. Over the years, different types of tents have been developed to meet various requirements of the military, explorers, nomads, etc. These tents are made of cotton canvas with wax emulsion treatment to provide water repellency. The development of high strength, rot proof, hydrophobic synthetic fibers along with improved polymer coating has given an impetus in the use of coated fabric as a membrane material to envelope very large building structures of the size of stadia or airports.

The fabric envelope is capable of resisting the elements of weather such as wind, rain, snow, sunlight, and even biological degradation. Although lower cost transparent film can be used as a membrane material, it lacks durability. Coated fabrics are the most widely used material because of high strength and environmental resistance. The advantages of these fabric envelope buildings are summarized below [20,21].

- The coated fabric envelope is much lighter than conventional building material (may be 1/30th) requiring much less structural support and reinforcements. This reduces the cost of the building.

- It provides large obstruction-free spans, suitable for large gatherings.
- The construction time is much shorter.
- Smaller envelopes can be dismantled and reerected elsewhere.
- Fabric envelopes are better resistant to natural hazards like earthquakes.

7.2.1 MATERIALS

The fiber and the fabric requirements of the coated fabric envelope are quite stringent. They should have the following properties:

a. High strength, to withstand the tension applied during construction of the structure, weight of the suspended fabric, and stresses due to wind, rain, and snow

b. The fiber should be creep resistant and have high modulus, for dimensional stability and resistance to deformation.

c. Retention of mechanical property in widely varying temperature conditions

d. Resistant to water, sunlight, and atmospheric pollutants

e. Long life

f. Low cost

The cost factor eliminates high-performance fibers such as aramids and carbon fibers. The choice is thus restricted to high tenacity polyester and glass fiber, out of which polyester is more popular. Continuous filament yarns are preferred over staple yarns because of higher strength and resistance to extension. The twist level is kept low to prevent fiber slippage and yarn rupture. Usually, woven structures are preferred for rigidity and dimensional stability. The type of weave should be such as to produce good yarn packing to minimize fabric deformation under tension and to provide a certain level of resistance to water and wind penetration. Generally, plain weave and 2×2 basket weaves are used. The high tenacity polyester fabrics used for architectural purposes have been categorized into a few fabric types, with fabric weight ranging from 220 to 630 g/m^2. A typical fabric is made of 1000 d yarn in plain weave (9.5 $\times$ 9.5 ends $\times$ picks/cm) having a weight of 220 g/m^2 [20,22].

The coated fabrics are orthogonally anistropic, i.e., they elongate differently in warp and weft directions. This aspect is to be taken care of in designing textile structures. Weft inserted warp knit and multiaxial knits have been tried to achieve isotropic structures, but they have not been successful, because on coating, the former develops anisotropy, and the latter results in a very stiff material [22].

Glass fabric has high strength, resists stretching, does not wrinkle, does not burn, and has high reflectivity, keeping the interior of the structure cool.

The polymer used for coating or laminating, the architectural fabric, should impart certain important properties to the membrane fabric. These properties are as follows [20,23].

- waterproofness
- impermeability to air
- resistance to abrasion and mechanical damage
- resistance to weathering and pollution
- ability to transmit and reflect light (It should have adequate translucency to provide natural illumination in daylight hours.)
- weldability
- flame retardance

Such requirements are met by two polymers, they are PVC and polytetrafluoroethylene (PTFE). Coating is preferred to lamination because laminated fabrics tend to delaminate on repeated flexing and wind lashing. Because of the high sintering temperature of the polymer, PTFE can only be applied on glass fabric. Two materials have found success as membrane material for envelope, PVC-coated polyester and PTFE-coated glass fabric [20,21,24]. A third material, silicone-coated glass fabric, is also emerging as a material of choice.

PVC-coated fabrics exhibit a dirt pick up problem, which is usually minimized by applying a thin top coating of polyurethane or acrylic lacquer. White is considered the color of choice, because the high reflectivity reduces surface temperature and enhances service life [3]. Translucency of the fabric is achieved by controlling the yarn density and by properly formulating the PVC compound. PVC-coated fabrics are easily joined by welding and are easier to handle for construction of the structure. A disadvantage of PVC is its slow embrittlement due to gradual loss of plasticizer. However, a properly designed fabric has a life of over 15 years. Different grades of these fabrics are available in the trade to meet the varying load-bearing capacities. The fabric weight ranges from 600 to 1000 g/m^2. These fabrics, because of their inherent flexibility, are suitable for semipermanent structures that may be taken down and reerected on a new site [20,22,25].

PTFE-coated glass fabric has several advantages. The polymer is chemically inert, self-cleaning, highly resistant to weathering, inherently translucent, and has excellent flame retardant properties. There are, however, certain problems in using this fabric. The abrasion resistance of PTFE is poor, the fabric is brittle causing handling problems, and joining of the fabric panels is difficult. The abrasion resistance is improved by incorporating glass filler in the outermost layer of the coating. The brittleness of glass fabric is reduced by a finishing treatment. Hot-melt PTFE resin is used for joining the fabric panels. Different grades of fabric are available with fabric weight ranging from 1250 to 1450 g/m^2. The fabrics have a solar reflectance of about 70% and transmission of about

15%. In view of the durability and stiffness of the fabric, they are used for permanent structures, that have a life of over 20 years [23].

7.2.2 STRUCTURES

Structures using coated fabric envelopes can be classified into three main types. They are tents, air inflated structures, and tensile structures. As stated earlier, wax-emulsion-treated canvas has been in use as tent cover for a long time, however, they have now been replaced by PVC-coated polyester. In a tent, the fabric is draped on a frame and is not tensioned. In large structures, the fabric is tensioned by air inflation or cables [20].

7.2.2.1 Air Supported Structures and Shelters

In these structures, the fabric assembly that serves as the roof is anchored and sealed to a ground foundation. Air is pumped inside to inflate and tension the envelope (Figure 7.5). The pressure required is only 3% above the ambient, therefore, it does not affect the comfort of the occupants. The entrance and exit doors are air locked to minimize drop of pressure when opened. Any fall of pressure is automatically corrected by pumping in air by a compressor. A double-layer roof is used to provide thermal insulation. Air houses were initially developed for protection of radar antennae and telecommunication equipment from high velocity winds and weather. The material used for radomes are neoprene- or hypalon-coated glass for transparency to electromagnetic

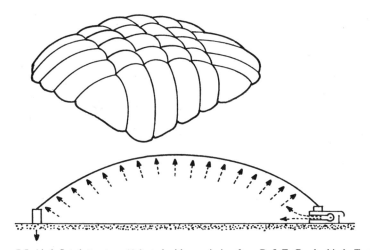

Figure 7.5 Air-inflated structure. (Adapted with permission from R. J. E. Cumberbirch. *Textiles,* Vol. 16, no. 2. 1987 © Shirley Institute, U.K. [21].)

radiation. Air supported buildings are now used for sports halls, exhibition halls, mobile field hospitals, storage bases, swimming pools, etc. For these purposes, the fabric widely used is PVC-coated polyester. These types of buildings are very stable, because highly flexible fabric distorts to distribute the damaging load due to strong winds and returns to the original shape after the wind has abated. The response to wind buffeting can be changed by varying the internal pressure using an automatic control system related to wind speed [21,24,25].

Another way of supporting the membrane structure is to use air beams. Air beams are convex, air-inflated (30–70 kPa) support tubes, which can be up to 1 m in diameter. They are mainly used by the military as lightweight shelters [20].

7.2.2.2 Flexible Barrier Storage System

Defense equipment and weapon systems, if left unprotected from weather, while not in operation, may undergo corrosion and microbiological degradation due to uncontrolled humidity and condensation of moisture due to temperature fluctuation. This may lead to operational failure and delays in reactivation of the equipment. Preservation by surface coatings such as paint and grease is not a long-time solution. After extensive experimentation, it has been found that the equipment could be preserved indefinitely in a controlled humidity between 30–40% relative humidity.

Flexible barrier storage systems are used by defense forces for preserving military equipment like tanks, helicopters, aircraft, and weapon systems from corrosion, rot, mildew, insects, dust, pollutants, and UV degradation. The equipment is mounted on a baseboard and is completely covered by reusable flexible barrier or shroud. Proper sealing is done by zip fasteners and Velcro to make the enclosure airtight. The cover is connected to a portable, solid desiccant-type dehumidifier through a flexible duct. Dehumidified air is circulated in the enclosure to maintain a relative humidity between 30–40%. There is no pressure difference between the inside and the outside of the envelope. Inspection windows and visual detectors are provided to monitor the relative humidity inside. An alarm system is also available in case the RH alters from the desired value. Several hardware systems can be connected to a single dehumidifier by a manifold. With this system, the equipment can be preserved for very long periods and can be reactivated in a very short time, enhancing the operational efficiency. Another major advantage is the flexibility of choice of storage sites. An alternative to the dehumidification process is to evacuate the enclosed envelope. However, with the vacuum system, it is difficult to maintain 30% RH continuously. Besides, the barrier material clings to the object forming creases leading to rupture and air leakage. A flexible barrier system showing protection of a tank is shown in Figure 7.6.

The barrier material should be lightweight, strong enough to resist environmental stresses, waterproof, flame retardant, and should have low water vapor

Figure 7.6 Flexible barrier system for protection of tank. Courtesy M/S Arctic India Sales, Delhi, India.

permeability, below 1.2 g/24 h/m^2. Different materials are used in the trade, they are butyl and PVC-coated nylon or polyester fabric. However, calendered multilayered PVC film is also used. The life of a shroud is between 7–9 years.

7.2.2.3 Tension Structures

In these structures, metal pylons or frames and tension cables are used to support the fabric. The fabric is tensioned by cables attached to the fabric by clamps. The tensioned structures are curvilinear and may be paraboloid or hyperbolic-paraboloid in shape. The curvature and prestress due to tensioning resists externally applied loads. To maintain rigidity and stability of the structure, multidimensional tensioning with properly designed curvature is required. These structures are used for permanent buildings and use generally PTFE-coated glass fabric [20,21,24].

Major challenges in the construction of these structures are their rigidity and stability to high velocity winds, rain, and snow. The designing, therefore, requires a knowledge of textile, engineering, and architecture. There are about 150 structures all over the world, which have been erected within the last two decades. These include airport terminals, stadia, and department stores. The largest being the Haj terminal building at Jeddah International Airport for which about 500,000 m^2 of PTFE glass fabric was used. These buildings have a distinct structural identity, the diffuse natural light through the roof is gentle to the eye

TABLE 7.1. Properties of Awning Fabrics.

	Types			
Properties	Vinyl-Laminated or Vinyl-Coated Polyester Backlit	Vinyl Coated Cotton	Acrylic-Painted Polyester/Cotton	100% Acrylic Woven
1. Base fabric	Polyester	Cotton	Polyester/cotton	Solution dyed acrylic or modacrylic
2. Coating / finish	Vinyl/acrylic	Vinyl	Acrylic	Fluorochemical finish
3. Wt, g/m²	540–740	500	400–450	300–330
4. Width, cm	155–180	78	78	155
5. Colors, top	Several solids	Several solids and stripes	Several solids and stripes	Several solids and stripes
6. Colors, underside	White/clear	Pearl gray	Pearl gray	Same as top
7. Opacity	Translucent	Opaque	Opaque	Opaque
8. Backlit translucency	Yes	No	No	—
9. Durability, years	5–10	5	5	5
10. UV resistance	Yes	Yes	Yes	Yes
11. Mildew resistance	Yes	—	—	Yes
12. Water repellency	Yes	Yes	Yes	Yes
13. Flame retardancy	Yes	Yes	No	Modacrylic—yes Acrylic—no

Compiled from Reference [26].

and gives a delightful ambience. The high translucency and strong membranes could one day permit games like soccer to be played under cover on real turf. Development work is going on to construct membrane structures to aid crop production in desert regions [20,24].

7.2.3 AWNINGS AND CANOPIES

These are architectural projections of a building. They provide shade, weather protection, decoration, and a distinct identity to the building. Awnings are wholly supported by the building to which they are attached by a lightweight rigid or retractable frame over which awning fabric is attached. Canopies are supported from the building as well as from ground. Illuminated or backlit awnings offer high visibility to commercial buildings. This is achieved by attaching lighting to the frame beneath the fabric cover. The awning and canopy fabrics are similar, however, backlit awnings are more translucent. Some of the desirable properties of awning fabrics are as follows [20]:

- resistance to ultraviolet radiation
- flame retardancy
- mildew resistance
- cleanability

From the various types of commercially available awnings listed in the "Awning fabric specifiers guide," it is seen that the major types of fabrics used for awning and canopies are as follows:

- vinyl-laminated or vinyl-coated polyester-backlighting
- vinyl-coated cotton
- acrylic-painted cotton or polyester cotton
- 100% acrylic woven

Out of these, vinyl-laminated or vinyl-coated polyester-backlighting is the most popular. Some important features of these classes are given in Table 7.1.

7.3 FLUID CONTAINERS

The low permeability of certain polymeric coatings have made coated fabrics a material of choice, as lightweight flexible containers for both gases and liquids. The containers for gases are known as inflatables and are meant to provide buoyancy. The liquid containers are used as storage vessels for fuels, water, etc. [27].

7.3.1 INFLATABLES

These are designed to contain air or carbon dioxide and are meant for buoyancy applications. One of the important uses of inflatables are in lifesaving aids such as in life jackets and in emergency rafts. Life jackets can be inflated by air, or automatically by carbon dioxide cylinder in case of an emergency. The buoyancy is properly designed so that the wearer is held in correct position in water. Modern life rafts are designed with capacities of 40 persons. They have tubular multiple buoyancy chambers, so that they remain afloat even if accidentally punctured. Emergency life rafts are inflated in seconds, by solid/liquefied carbon dioxide. All the rafts are fitted with canopies for protection against weather. The canopy usually has a fluorescent orange color for easy detection. The floor of the raft should be watertight and provide insulation from seawater. These rafts are made from nylon fabric (130 g/m^2), coated on both sides with polyurethane. Some naval rafts use nylon laminates containing an intermediate layer of butyl rubber, which has low permeability to gases [27,28].

Inflatable crafts are used also for patrol duty by the Coast Guard or for leisure. They are inflated by air at pressure between 15 to 25 kPa and may be propelled by oars or a small overboard motor (Figure 7.7). The air cushion of a hovercraft is also contained in a coated fabric skirt fitted around the hull at a pressure of 3 kPa. The lift to the craft is propelled by a downward air flow.

An interesting application of an inflatable is for constructing temporary bridges for ferrying heavy military vehicles across a river or pond. This is

Figure 7.7 A motorized raft. Courtesy M/S Swastik Rubber, Pune, India.

done by laying a set of floats across the river. The floats are air inflatable buoys, containing tubular, multiple buoyancy chambers made of coated fabric. A typical float fabric is a flexible composite, consisting of two layers of nylon fabric with neoprene coating in the intermediate layer as well as on both outer layers. The weight of base fabric and coated fabrics are $\sim$240 g/m^2 and $\sim$1500 g/m^2, respectively. Other applications of inflatables include oil booms to contain oil slicks in the sea and air bags for manipulation of awkward heavy objects over water in offshore industries. Deflated air bags are used in the salvage operation of sunken objects, like crashed aircraft from sea by inflation [27,28].

Hot air balloons for leisure are also inflatables that are made of lightweight nylon fabric (30–60 g/m^2) in rip-stop construction, coated with a thin layer of PU to reduce porosity. However, due to the temperature of hot air inside ($\sim$100°C), UV exposure and manhandling, the fabric can last for only 300–500 flight hours [28].

The main components of inflatables are (a) coated fabrics, usually calendered; (b) the adhesive; and (c) the inflation valve. The factors that are to be considered for designing inflatables, particularly crafts for marine applications, are as follows [29]:

(1) Strength of the coated fabric: the inflatables may be considered as a thin-walled cylinder. The strength requirement of the inflatable is obtained from the hoop stress, exerted due to pressure of inflation; i.e., $pD/2t$ (where $p =$ pressure, $D =$ diameter of the tube, and $t =$ thickness), along with a suitable safety factor. Usual inflation pressure is about 15–20 kPa. The strength of the coated fabric, tensile and tear, is obtained from the textile substrate and depends on type of yarn and construction. Normally, nylon or polyester woven fabrics are used.

(2) The polymer coating: this is dependent on the type of use and life expected. The polymers generally used are neoprene, hypalon, and polyurethanes.

(3) Airtightness: the polymeric coating should be pinhole-free for proper gas holding properties.

(4) Resistance to weathering and UV degradation: this is required as the inflatables are used outdoors for very long periods.

(5) Resistance to abrasion: a higher resistance to abrasion is required for protection against damage due to launches and landings on sandy and rocky shorelines.

(6) Resistance to oil and seawater: these are important requirements as the crafts are used in marine and oily environments.

(7) Flexibility: the fabric should retain its flexibility over a wide range of temperatures (-30 to $40°C$). Low temperature flexibility down to $-45°C$ is required for rafts inflated by solid carbon dioxide.

The coated fabrics for inflatables are produced by calendering or spreading. However, calendering is more suitable for compact, thick polymeric coatings on heavy-duty base fabric. The coated fabric, which usually has a width of ~150 cm, is cut into panels of various shapes depending upon the shape of the finished article. The edges of the fabric to be adhered are buffed and joined by a cold curing adhesive to form the inflatable structure. The adhesive should have certain important characteristics. They are compatibility with the polymer, good green tack, rapid cure, resistance to oil and water, and appropriate cure strength.

7.3.2 LIQUID CONTAINERS

Coated fabrics are particularly suitable for collapsible storage containers and for transporting liquids by land and sea. Dracone barges, which are large flexible containers, are suitable for towing liquids on the ocean. One of the latest applications of these are in pollution control due to oil spillage in the sea. The spilled oil is contained by a boom system placed around the slick. The oil is then pumped into dracones and transported away for reclamation or incineration. Dracones can also be used to transport the large quantity of detergent required to disperse the oil spill [27]. Collapsible containers are used as fuel tanks for military aircraft and as fuel containers in temporary air fields. The coated fabric can be heavy, about 5000 g/m^2 on a nylon base fabric of about 600 g/m^2. The polymer used is generally neoprene. However, for storage of fuels, nitrile rubber is used, and an approved grade of PU is used for storage of drinking water [28].

A detailed investigation has been reported to find out the parameters of weave construction by which tear resistance of base fabric used for large collapsible fuel tanks (~200 kl) could be maximized. High-tenacity nylons of 840 d and 1050 d were woven into plain, twill, and basket weave designs with different yarn densities. Considering both tear resistance and dimensional stability, a 11 × 11 ep/pp cm, 2 × 2 basket-weave fabric, constructed with 1050 d yarn gave best overall performance [30].

7.4 TARPAULINS

Tarpaulins are used as covers to protect commodities from damages due to weather. In the agricultural sector, they are extensively used for protecting grain and machinery, while in the construction industry, they are used to protect building supplies like timber and wet concrete. In transportation of goods by road, they are used to cover the cargo. Traditionally, canvas covers made from heavy-duty cotton fabrics with a wax emulsion and a rot-proof treatment have been used for this purpose. However, canvas covers have now been completely replaced by 450–500 g/m^2 PVC-coated fabric because of their lighter weight and inherently water- and rot-proof nature. Some important properties of tarpaulins are (1) waterproofness; (2) strength; (3) tear, puncture, and abrasion resistance; (4) flexibility at a wide range of temperatures; and (5) durability.

The coated fabrics used are vinyl-coated nylon or polyester fabric with weights between 500–600 g/m^2 ($\sim$350 g/m^2 coating). For covers likely to be contaminated by fuels/oils, neoprene, hypalon, or PU-coated fabrics are used [31]. High density polyethylene woven fabrics laminated on both sides by low density polyethylene films are also being used as covers. For military use, tarpaulins should have camouflage properties as well. A desirable feature for vehicle covers is reversibility that would cater to change of terrain, e.g., from green belt to desert terrain.

7.5 AUTOMOTIVE AIR BAG FABRICS

Untill recently, safety belts have been the only protection for passengers in a car crash. During the last decade, air bags or inflatable restraints have gained significant importance as protection for the driver and passengers in case of a collision. The original bag was designed for head-on collision, but at present, side impact bags, knee bolsters, side curtain, etc., are available for safety in any type of crash. Because frontal collisions are a major cause of accidental deaths, air bags are being introduced as a standard item in vehicles by legislation in the U.S. Consumer consciousness coupled with legislation has resulted in rapid growth of air bags during the last decade [20,32,33].

The air bag is built into the steering wheel and the instrument panel of the car. An air bag module consists of the air bag, crash sensors, and mounting hardware. In case of frontal collision equivalent to 20 km/h against a wall, sensors set off the igintor of the inflator. Pellets of sodium azide in the inflator ignite and release hot nitrogen gas. The gas passes through a filter to remove ash and other particulate matter and inflates the air bag. A pressure of 35–70 kPa is generated. The air bag is fully inflated within 60 ms and cushions the occupant from impact. After absorbing the forward force, the air bag deflates after 120 ms [20]. The capacity of a driver-side air bag is normally $\sim$65 L, but for smaller cars, it is $\sim$35 L. Passenger-side air bags are much larger (100–300 L) [33].

The air bag fabric is made from nylon 66 because of its high weight-to-strength ratio and is preferred over polyester because of higher elongation, allowing the force to distribute widely. The driver-side air bag has an elastomeric coating to provide heat shielding and ablative protection to the fabric from the hot gases. Moreover, coating seals the fabric pores and permits precise control in the deployment of the air bag.

The coated fabric should be antiblocking, have high tensile and tear strength, good adhesion, and long-term flexibility to cyclic temperature changes between extreme cold and hot conditions (-10 to $120°C$). Besides, the fabric should

TABLE 7.2. Neoprene-Coated Fabrics for Air Bags.*

Nylon Yarn d	Construction Plain Weave Ends/cmx Picks/cm	Base Fabric, wt. g/m^2	Coated fabric, wt. g/m^2
840	10 × 10	190	280
420	18.4 × 18.4	185	260

* Adapted with permission from E. T. Crouch, *Journal of Coated Fabrics,* vol. 23, Jan. 1994. © Technomic Publishing Co., Inc. [32].

be soft and smooth so as not to cause secondary abrasion or bruises and have good packageability. Untill recently, two types of neoprene-coated fabrics were used because of better environmental stability and flame retardant properties of neoprene [32]. One is a heavy fabric made of 840 d nylon and the other a lighter fabric woven from 420 d nylon. The details are given in Table 7.2.

The need to enhance the life of the air bag and further reduce the size led to the development of silicone-coated air bags. Silicones are chemically inert and maintain their properties for a long time at temperature extremes. An aging study of both neoprene- and silicone-coated fabrics was carried out at 120°C for 42 days. The elongation of the fabrics prior to aging were about 40%. After aging, the elongation of silicone-coated fabric was 32% but that of neoprene-coated fabric dropped sharply to only 8%. This has been attributed to the poor compatibility of neoprene with nylon. It is possible that chlorine in the neoprene produces an acidic environment, embrittling the nylon fabric. Another drawback of neoprene-coated material is that it should be dusted with talc to prevent self-adhesion, which creates dust in the vehicle interior following deployment of the bag. Silicone-coated fabrics are more flexible and abrasion resistant than neoprene-coated ones. Moreover, because of better durability and compatibility of silicones with nylon, a thinner coating is adequate. A silicone-coated air bag fabric made of 420 d/315 d nylon weighs only about 200 g/m^2. They can, therefore, be packed in smaller modules [32].

For this reason, silicone coating is rapidly replacing neoprene coating for driver-side air bags. Passenger-side air bags are made of uncoated nylon fabrics as the functional requirements are not that stringent. It is estimated that by the end of the decade, the requirement of coated air bag fabric will be between 50–75 million sq. meters, making it one of the most important growth sectors for coated fabrics both in volume and in value terms [20].

7.6 CARPET BACKING

A variety of fibers, both man-made and synthetic, are used as face fabric of carpets. The commonly used ones are cotton, wool, rayon, polyester, polypropylene, and nylon. The carpets are given a backcoating or a fabric backing to impart strength and durability. Three types of carpet backing are used in industry [34,35].

(1) *Secondary-backed carpet:* in these carpets, a secondary backing fabric, usually jute or polypropylene, is bonded to the back of the carpet by an adhesive.

(2) *Unitary coating:* this consists of a simple application of an adhesive layer on the back of the carpet without any secondary backing.

(3) *Foam backing:* the back of the carpet consists of a thin cushion of foam as its integral part.

The backcoating process imparts certain important properties to the carpet, viz., tuft binding, dimensional stability, resistance to water, reduced pilling, resistance to edge fraying, etc. In order to achieve all these properties and to obtain adequate adhesion with the secondary backing, it is important to select the right adhesive.

Various materials have been used as adhesive over the years. They are natural rubber latex, SBR latex, EVA emulsion, PVA emulsion, starches, etc. Out of these, SBR latex and carboxylated SBR latex are the most widely used adhesives. SBR latex is obtained by emulsion polymerization process. For backcoating purposes, the SBR latex is formulated with certain additives. They are calcium carbonate (extender), surfactant (frothing aids), and polyacrylate thickeners for adjusting the viscosity [36].

The tuft locking and stiffness depends on the coating weight of the latex as well as on the formulation such as filler content and styrene content of the latex. The stiffness and tuft locking increases with styrene content and the add on. Increase in filler content decreases the tuft locking but increases stiffness [34].

Secondary-backed carpet is a type of carpet mainly used for residential purposes. Different methods are used for lamination of the backing jute fabric. These consist of applying an undercoat of SBR latex on the back side of the carpet and an adhesive coat on the jute surface and bonding the two fabrics in between laminating rolls. The undercoat is usually frothed by air, for better weight control, has higher viscosity (14,000–18,000 cps) and higher extender content (400–800 parts/100 parts latex). The adhesive coat has lower viscosity (9000–10,000 cps), lower extender content (300–400 parts/100 parts latex), and is unfrothed [35,36].

An important method of lamination is the pan coat, jute coat process, where the undercoat and adhesive coats are applied from pans to the carpet back and jute surface, respectively, by kiss-coating or roller-coating techniques. The coated surfaces are then bonded between press rolls and cured. In the direct-coating system which is more popular (Figure 7.8), the undercoat (frothed SBR

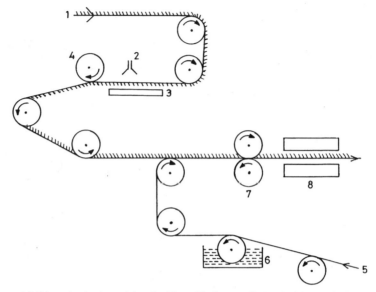

Figure 7.8 Direct lamination of jute backing: (1) Carpet, (2) Frothed latex, (3) Bed plate, (4) Metering roll, (5) Jute fabric, (6) Adhesive, (7) Press rolls, and (8) Oven.

latex) is spread directly on the carpet back using a bed plate and a doctor roll; an adhesive coat is applied on jute surface by roller coat from pan, and the two surfaces are bonded in the usual manner [34,35].

In the unitary backing process, a layer of adhesive, usually SBR latex, is applied on the back side of the carpet by knife on roll, kiss coating, or roller coating followed by curing. As mentioned earlier, no secondary backing is used. The latex used has lower extender content (100–150 parts/100 parts of latex) and lower viscosity (7000–9000 cps). The process gives better tuft lock and dimensional stability. These carpets are used where the traffic requirement is high [34,36].

For foam backing [37], a precoat of SBR latex is first applied on the back of the carpet and dried. A foam coat is next applied by blade or roller coating. The coated fabric is then dried and vulcanized in a stenter.

Different types of foam systems based on air-frothed SBR or SBR-NR latex blend are used. In the chemical gelation system, coagulation occurs due to destabilizing the colloid, forming a rubbery continuum. Two types of gelling agents are commonly used: sodium silicofluoride, which operates at room temperature with delayed action, and ammonium salt-zinc oxide system, a heat-sensitive gellant. In a nongel foam process which is also quite popular, carboxylated latex is used along with water-soluble melamine-formaldehyde (MF) resin. Cross-linking occurs via the carboxyl groups of the latex with MF resin. The choice of the foam system depends on the technical requirements, simplicity of the process, and cost.

7.7 TEXTILE FOAM LAMINATES FOR AUTOMOTIVE INTERIORS

Textile fabrics are gradually replacing vinyls for car seats and interiors because of their soft handle, design color, and pattern options. The fabrics used are generally pile fabrics, of nylon or polyester. The fabric is converted into a trilaminate, comprised of face fabric, a polyurethane foam of 2–10 mm thickness, and a lining fabric. The face fabric provides an attractive look, the foam provides a soft cushioning effect, and the liner prevents the foam from wear. The most common method of lamination is flame lamination. However, laminating with dry and hot-melt adhesives is emerging as an alternative. The different methods used for lamination are discussed below [38].

7.7.1 FLAME LAMINATION

In this process, a roll of polyurethane foam passes over an open flame, resulting in melting of the surface of the foam, which then functions as the adhesive. This material is then bonded to fabric, by passage through nip of the laminating roll (Figure 7.9). The process is repeated for laminating the liner, on the back side of the foam.

The process is simple, does not require an oven, and gives fast line speeds. The burning of the foam, however, liberates toxic gases such as hydrogen chloride and cyanides, which have to be properly ventilated. The flame length has to be properly adjusted as large flame burns too much of the foam, while a small flame leads to insufficient melt and poor bonding. The process is restricted to foams that can be melted.

7.7.2 DRY ADHESIVES

Lamination can be done by scatter coating dry powder or by film adhesion of thermoplastic polymer. In scatter coating, powders of 20–200 μm of

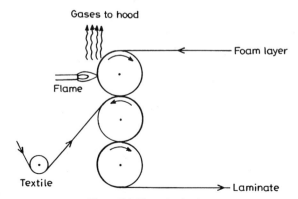

Figure 7.9 Flame lamination.

polyester, polyamides, or EVA obtained by cryogenic grinding of polymer granules are scattered on the foam substrate. Foam containing the powder then passes through heaters to activate the adhesive by melting. The textile web is also heated to near the softening temperature of the adhesive. The lamination of the foam and textile layers is carried out by passage through nip of the laminating rolls. In film adhesion, a film adhesive obtained in roll, by melt extrusion is placed on the foam substrate, and the two layers are passed through heaters to melt the adhesive. Heated textile substrate is then laminated to the foam layer in a lamination station. The laminates are cooled prior to winding. These methods have the advantage in that no emissions are produced. They, however, require large ovens and are unsuitable for temperature-sensitive fabrics.

Adhesives can be applied on the substrate for lamination by rotogravure or by spraying processes, but the viscosity and pot life of the adhesive are constraints of their applicability.

7.8 FLOCKING

Flocking is the application of short fiber (flock) on an adhesive-coated substrate in vertical position. The substrate may be woven or knitted textiles, leather, paper, polymer film, etc. Flocking is used for producing a variety of items with aesthetic appeal, such as draperies, bedsteads, carpets, and artificial fur and suede. The flocking process provides an economical means of production of pile texture.

The flock commonly used in trade are nylon or viscose, 1640 d, of 0.1 to 0.6 mm length. For production of the flock, tows of fibers are extruded through the nip of rotating rolls and are cut by a blade. The speed of the rolls determines the flock length. The flock is then thoroughly washed, activated by cationic detergent, and graded prior to electrostatic deposition on the substrate. The textile substrate used are mainly woven or knitted cotton/viscose fabric. The adhesives are usually latices of NR, SBR, or aqueous dispersions of acrylics or polyurethanes. The viscosity of the adhesive is adjusted using polyacrylate thickeners to prevent strike through of the adhesive during coating.

The flocking process consists of coating the textile substrate by knife on roll, rotary screen printing, or gravure roll coating for repeat patterns. Flock is then applied on the substrate electrostatically. For this purpose, the flock fibers are introduced into a high voltage field as a result of which, as charge carriers, the fibers are transported to the adhesive-coated substrate at right angles. Adhesive-coated substrate forms the earth pole and is, in addition, vibrated by a set of rollers. The fibers that penetrate the adhesive layer are retained there, forming a dense pile. Excess flock is removed by suction, and the product is passed through a tunnel drier, which is circulated by hot air at high temperature. The product is cooled, excess flock is removed by brushing followed by suction,

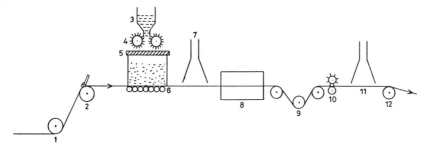

Figure 7.10 The flocking process: (1) fabric unwind, (2) coating of adhesive, (3) hopper containing, (4) rotating brush, (5) electrostatic depositer, (6) vibrating rolls, (7) suction, (8) drying oven, (9) cooling rolls, (10) brushing, (11) suction, and (12) winding.

and it is cut into suitable lengths and packed. A schematic layout of the process is given in Figure 7.10.

7.9 REFERENCES

1. G. R. Lomax, *Journal of Coated Fabrics,* vol. 15, Oct., 1985, pp. 127–144.
2. D. Morley, *Journal of Coated Fabrics,* vol. 14, July, 1984, pp. 46–52.
3. *Textiles,* vol. 10, no. 3, 1981, pp. 65–68.
4. *Poly Vinyl Chloride,* H. A. Sarvetnick, Van Nostrand Reinhold, New York, 1969.
5. *Polyurethane Handbook,* G. Oertel, Hanser Publishers, Munich, 1985.
6. B. L. Barden, *Journal of Coated Fabrics,* vol. 24, July, 1994, pp. 10–19.
7. *Kirk Othmer Encyclopedia of Chemical Technology,* 3rd Ed., vol. 14, 1979, pp. 231–249; 4th Ed., vol. 15, 1995, pp. 177–192, both John Wiley and Sons, New York.
8. Fibrous composite poromerics, W. Reiss, in *Coated Fabric Technology,* vol. 2, Technomic Publishing Co., Inc., Lancaster, PA, 1979, pp.45–56.
9. P. Durst, *Journal of Coated Fabrics,* vol. 13, Jan., 1984, pp. 175–183.
10. J. Hemmerich, J. Fikkert and M. Berg, *Journal of Coated Fabrics,* vol. 22, April, 1993, pp. 268–278.
11. B. Zorn, *Journal of Coated Fabrics,* vol. 13, Jan., 1984, pp. 166–174.
12. R. Markle and W. Tackenberg, *Journal of Coated Fabrics,* vol. 13, April, 1984, pp. 228–238.
13. Man made leather, O. Fukushima, in *Coated Fabric Technology,* vol. 2, Technomic Publishing Co., Inc., Lancaster, PA, 1979, pp. 119–131.
14. C. Chu, Z. Mao and H. Yan, *Journal of Coated Fabrics,* vol. 24, April, 1995, pp. 298–312.
15. B. Kunst and S. Sourirajan, *Journal of Applied Polymer Science,* vol. 14, 1970, pp. 1983–1996.
16. H. Strathman, P. Schieble and R. W. Baker, *Journal of Applied Polymer Science,* vol. 15, 1971, pp. 811–825.

17. C. Lemoyne, C. Friedrich, J. L Halary, C. Noel and L. Monnerie, *Journal of Applied Polymer Science,* vol. 25, 1981, pp. 1883–1913.
18. Absorptive vinyls, B. M. Murphy, in *Coated Fabric Technology,* vol. 2, Technomic Publishing Co., Inc., Lancaster, PA, 1979, pp. 103–112.
19. J. Duncan, *Journal of Coated Fabrics,* vol. 13, Jan., 1984, pp. 161–165.
20. *Handbook of Industrial Textiles,* S. Adanur, Ed., Technomic Publishing Co., Inc., Lancaster, PA, 1995.
21. R. J. E. Cumberbirch, *Textiles,* vol. 16, no. 2, 1987, pp. 46–49.
22. H. Mewes, *Journal of Coated Fabrics,* vol. 22, Jan., 1993, pp. 188–212.
23. Architectural Fabrics, M. Dery, in *Coating Technology Handbook,* D. Satas, Ed., Marcel Dekker, New York, 1991.
24. B. Foster, *Journal of Coated Fabrics,* vol. 15, July, 1985, pp. 25–39.
25. G. R. Lomax, *Journal of Coated Fabrics,* vol. 15, Oct., 1985, pp. 127–144.
26. *Awning and Canopy Fabric Specifiers Guide,* Industrial Fabric Association International, 1996, 73/6, pp. 37–48.
27. K. L. Floyd, *Textiles,* vol. 6, no. 3, 1977, pp. 78–83.
28. G. R. Lomax, *Journal of Coated Fabrics,* vol. 15, Oct., 1985, pp. 127–144.
29. E. Sowden, *Journal of Coated Fabrics,* vol. 13, April, 1984, pp. 250–257.
30. L. H. Olson, NTIS report no. DAAG 53-76-C-0141, 1977.
31. G. R. Lomax, *Textiles,* vol. 14, no. 2, 1985, pp. 47–56.
32. E. T. Crouch, *Journal of Coated Fabrics,* vol. 23, Jan., 1994, pp. 202–219.
33. F. A. Woodruff, *Journal of Coated Fabrics,* vol. 23, July, 1993, pp. 14–17.
34. R. L. Scott, *Journal of Coated Fabrics,* vol. 19, July, 1989, pp. 35–52.
35. K. Stamper, *Journal of Coated Fabrics,* vol. 25, April, 1996, pp. 257–267.
36. Non apparel coating, D. C. Harris, in *Coated and Laminated Fabrics New Processes and Products, AATCC Symposium Proceedings,* April 3–4, 1995, Denvers, MA, U.S.A., pp. 4–21.
37. Latex applications in carpets, D. Porter, in *Polymer Latices and Their Applications,* K. O. Calvert, Ed., Applied Science, London, 1982.
38. A comparative analysis of laminating automotive textile foam, J. Hopkins, in *Coated and Laminated Fabrics New Processes and Products, AATCC Symposium Proceedings,* April 3–4, 1995, Denvers, MA, U.S.A., pp. 250–267.

CHAPTER 8

High-Tech Applications

8.1 FABRICS FOR CHEMICAL PROTECTION[2]

AWARENESS of the risks and ill effects involved in working in an environment full of pervasive liquids and chemicals has necessitated the use of protective gear for employees in the workplace and persons in public places. In the developed world, the concerns related to the ill effects of toxic chemicals are much greater compared with developing and underdeveloped countries. Nevertheless, definite and distinct awakening is taking place, and new and strict laws are being promulgated to protect individuals and the environment from the menace of toxic chemicals.

The toxicity of chemicals in general depends upon their structure, physiological action, and mode of exposure [1]. It may be safely stated that each and every chemical known to date can be considered as toxic at some level of intake. However, the most toxic chemicals developed and stockpiled for use in chemical warfare may be categorized in four classes, namely, blood agents, choking agents, vesicants, and nerve agents. Out of the above classes, vesicants and nerve agents manifest their effect through skin absorption. Vesicants damage body tisssue and form painful blisters that are difficult to heal. Nerve agents are absorbed and transported through the bloodstream where they block the enzyme acetyl choline esterase that plays a key role in the neurotransmission cycle leading to incapacitation and fatality. Coated fabrics are used for protection against these two classes of agents. Some of the important characteristics of common chemical warfare (CW) agents of concern are summarized in Table 8.1.

The toxicity and hazards associated with other common chemicals used in industries are much below the chemicals listed in Table 8.1. Nevertheless, the workers in plants and industries likely to be exposed due to the nature of the job need to be protected from their ill effects.

[2]Contributed by V. S. Tripathi, DMSRDE, Kanpur, India.

TABLE 8.1. Properties of Important Chemical Warfare Agents.

Name of Agent	State at 25°C	Volatility at 25°C (mg/M^3)	Effect	Time Taken for Appearance of Symptoms
1. Pinacolyl methyl phosphono fluoridate (SOMAN)	Colorless liquid	3060	Nerve agent	Inhalation 1–5 min, skin 30–60 min.
2. Isopropyl methyl phoshono fluoridate (SARIN)	Colorless liquid	16400	-do-	-do-
3. Ethyl N,N'-dimethyl phosphoroamidecyanidate (TABUN)	Colorless liquid	516	-do-	-do-
4. o-Ethyl-S-2-diisopropylamino ethyl methyl phosphonothioate (VX)	Yellow liquid	16	-do-	-do-
5. bis-(2-Chloroethyl) sulphide (S Mustard)	-do-	930	Vesicant	3 hrs.

Coated fabrics play a key role in both civil and military applications as far as protection for the whole body is concerned. The whole gamut of coated fabrics used for protection of the human body may be conveniently classified in two broad categories, permeable or breathable and impermeable or nonbreathable. As the names suggest, the former allows free ingress and egress of air facilitating the dissipation of heat and evaporation of sweat, while the latter completely shields the wearer from the atmosphere. Obviously, the devices made of permeable-type fabrics can be used for longer duration of time due to comparatively low heat stress. However, for many applications where large quantities of toxic chemicals are handled or liquid splash may occur completely drenching the wearer, impermeable suits are preferred. Various types of permeable and impermeable coated fabrics used for protection against different chemicals and scenarios are discussed in this section.

8.1.1 TYPES OF CHEMICAL EXPOSURE RISKS

Chemicals can enter the human system through inhalation, ingestion, or absorption by the skin. The nature of the chemicals and the form and place of exposure decides the type of protection required. Some prominent scenarios include during war where chemical warfare agents may be used; emergencies involving accidental spills on highways; working with and handling hazardous waste; laboratory work; radioactive contamination; manufacturing operation for chemicals in pharmaceutical, electrical, and electronic industries; and in the handling of pesticides, insecticides, and herbicides. The United States Environmental Protection Agency (U.S. EPA) has classified the exposure scenarios and level of protection required in four broad categories. Coated fabrics are used for protection against skin absorption. Obviously, the use of protective clothing hampers the normal activities of the wearer, hence, a prudent

TABLE 8.2. Categorization of Exposure Scenarios and Implements Required for Protection.

Level	Example of Scenario	Protection Implements Required for
A	1. Production, storage, and packaging of extremely hazardous chemicals	Full face mask, suit completely encapsulating the body, gloves and overboots
	2. War field where toxic agents have been used (damages due to inhalation and skin absorption) due to dissemination of CW agents	
	3. Decontamination drill	
B.	1. Chemicals are highly toxic if inhaled but not absorbed through the skin	Self-contained breathing apparatus-(SCBA) Skin protection suit not completely sealed
	2. Atmosphere with less than 19.5% oxygen, e.g., fire, etc.	
	3. Splash of chemicals possible	
C.	1. Industrial contaminants, type and concentration known	Full face mask with appropriate canister having high efficiency particulate filter. Splash suit preventing direct contact of chemicals
	2. Particulate contaminants, e.g., radioactive dust	
D.	Nuisance contaminants with minimum hazards. No chemical immersion or splash	Standard work clothing and apron

selection of chemical protective garments based on the hazard and risk of exposure anticipated is very important. Table 8.2 gives an idea of possible scenarios and the recommended protective equipment required in the given scenarios.

It may be noted from Table 8.2 that the stringent requirements of level A hazards require complete encapsulation. This drastically reduces the time of use for chemical protective clothing (CPC) made of impermeable material. In civilian applications, it is possible to rotate the deployed manpower at short intervals to finish the task. However, in military applications, especially in war field rotation, it is not possible, hence, permeable/breathable suits are preferred. CPC requirements stipulated in B, C, and D levels of hazards are not critical, nevertheless, they should provide protection against chemical splashes in B and C levels of hazards.

8.1.2 MATERIALS FOR IMPERMEABLE CPC

Impermeable chemical protection clothing is normally made either by barrier coating on a fabric or films as such. The fabric used for coating or as a carrier of the barrier compound provides necessary strength to the clothing. Fabrics made of polyester, polyamide, and cotton and their blends are commonly used as carrier fabrics. In disposable limited-use garments, nonwoven fabric laminated with a barrier film is preferred. The materials used for civil and military applications (against CW agents) are different, and they are discussed below.

8.1.2.1 Civil Applications

Since the first use of chemicals for military purpose in World War I, tremendous advancement has taken place in understanding the barrier properties of polymeric materials vis-a-vis different types of chemicals in liquid or vapor forms. In the beginning of the 20th century, natural rubber was the only material available for coating. Prior to World War II, synthetic elastomers such as neoprene, polyvinyl chloride, and butyl rubber replaced natural rubber in civil and military CPCs. With the development of newer polymers like fluoropolymers, CPCs using a wide spectrum of elastomers, and thermoplastics were developed by leading manufacturers all over the world. Coextruded polymeric films have become a material of choice in limited-use disposable clothing. A list of different materials used in impermeable CPCs is given in Table 8.3 [2].

The criteria for selection of a material for a civil application depend on the permeation of the toxic vapor and penetration of the challenge liquid as evaluated by ASTM 739 and 903.

TABLE 8.3. **Common Barrier Materials in Impermeable CPCs.**

Elastomers—Unsupported/Reinforced
- butyl/bromobutyl
- chlorobutyl
- fluoroelastomers (Viton)
- urethane

Plastic Film Laminates—Coating
- chlorinated polyethylene
- PTFE
- polyethylene
- polyvinyl chloride
- polyvinylidene chloride

Bicomponent Constructions
- fluoroelastomer/butyl
- fluoroelastomer/neoprene
- PVDC/polyethylene
- neoprene/PVC

8.1.2.2 Military Applications

For military applications, the barrier properties are evaluated using S Mustard as the probe. This is considered the most penetrating of all the CW agents. A material is considered suitable for the protection of inanimate objects if it does not absorb more than 50% of the placed 1 μL of S Mustard on 1 cm^2 area of the material in 6 hrs. In personnel protection, vapor penetration should be very low because exposure to even 75 μg/m^3 concentration for one hour is enough to produce blisters. The test procedure of these materials is discussed in Chapter 9.

Two types of materials are used for fabrication of impermeable suits, etc.: (a) multilayer sandwich type and (b) coated nylon or polyester fabric [3]. In the former type, a barrier film of material such as polyvinylidene chloride, polyamide, or polyester is sandwiched between weldable polyolefin films. Sandwich layers are biaxially oriented before or after lamination for better mechanical strength. The overall thicknesses and weights of such multilayer films are 100 μms and 100–150 g/m^2, respectively. For heavy-duty application, such as adhoc collective protection, decontamination, and disposal of munitions, coated fabrics having much higher strength than multilayer films are preferred. Butyl or perfluorocarbon rubber is used for coating, and the weight of the fabric varies between 250–500 g/m^2.

8.1.3 MATERIALS FOR PERMEABLE CPCs

Permeable fabrics allow free passage of air, permitting sweat of the wearer to freely pass out as water vapor. The advantage gained in suits made of such fabrics is significant as the physiological load in the form of heat stress is much less compared to that of the impermeable suits. It is possible to indulge in light to medium work schedules wearing these suits for a sufficient length of time.

A layer of high surface area microporous carbon (pore width <20 Å) is impregnated/coated on different carriers for attenuation of challenge concentration of the chemical agents. The role of carbon in a breathable fabric is very critical, in that, it should preferentially adsorb chemical agents with minimal desorption during usage. High surface area microporous carbon performs this task remarkably well. Desorption from micropores where adsorption forces are enhanced due to proximity of walls in slit-shaped pores is comparatively difficult. Fortunately, such active carbons are very good adsorbents of chemical agents that are adsorbed through the skin at ambient temperature. Even a bed of 0.5–1 mm of microporous carbon available in NBC fabrics provides very good protection, provided the rate of adsorption is high. Different permeable fabric systems in use globally are described below along with their salient features [3,4].

8.1.3.1 Carbon-Impregnated Polyurethane Foam

Lightweight, low density, thin (1–2 mm) polyurethane appears to be good candidate matrix for impregnation of carbon. In fact, comparatively large quantities of carbon (200 g/m^2) can be impregnated in foam material, yet loss of carbon during work schedules and military operations is insignificant. However, the insulation properties of foam and its ability to adsorb fluids are serious disadvantages that cause a lot of physiological stress. Moreover, being voluminous material, the suits made of the carbon-impregnated foam cannot be packed in a small space. The deterioration of foam, especially in a hot and humid atmosphere, is another drawback. Suits of this type are currently being manufactured and used in France and some other countries.

8.1.3.2 Carbon-Impregnated Cotton Flannel and Nonwoven Fabric

Activated carbon in very fine powder form can be coated on open-structure carriers such as cotton flannel and nonwoven fabric using a suitable binder. In the case of flannel, air permeability is on the low side. This type of carbon-coated fabric is used in China for NBC suits. Thin wadding of nylon or polyester reinforced by open-structure cotton scrim is probably the best carrier for active carbon, and this type of coated fabric gives very good air permeability. Polychloroprene is used as a binder because of its flame retardant properties. It reduces the adsorbability of carbon only marginally. The carbon content per unit area (45–80 g/m^2) is about one-third compared to foam-impregnated material, nevertheless, a faster rate of adsorption due to the finer particle size and better air permeability more than compensates this shortcoming. Such suits are made and used in the U.K.

8.1.3.3 Bonded Spherical Carbon Adsorbents

In this system, microspheres of activated carbon having diameter 0.5–1.0 mm are point bonded on a carrier fabric. Approximately 150 g/m^2 carbon loading is achieved in this way, and this gives much better life when compared to carbon-coated fabrics. However, due to the larger granule size, the rate of adsorption is low, and larger quantities of carbon per unit area result in more weight, higher heat stress, and more cost. This type of microsphere-coated fabric is manufactured in Germany and the U.S. for military applications.

8.1.3.4 Active Carbon Fabric (Charcoal Cloth)

Activated surface area microporous adsorbent media in a fabric form was developed by Maggs [5] using viscose rayon fabric as precursor. The manufacturing process of activated carbon fabric comprises of pretreatment of the precursor fabric with a Lewis acid solution and the carbonization and activation in a carbon dioxide atmosphere. The surface properties of the end product can be controlled by careful selection of operating parameters [6]. This type of adsorbent media can be used in protective clothing after proper lamination with woven or nonwoven fabric. The quantity of activated carbon available per unit area of fabric is quite high (100–120 g/m^2), and the rate of adsorption is also very high. In spite of all the advantages, poor mechanical strength of charcoal cloth has found limited use in chemical protective clothing.

8.2 THERMOCHROMIC FABRICS

A number of high technology fibers have been deveoped in recent years, some of them are based on microencapsulation technology [7,8]. These include fabrics that release perfume on rupture of the microcapsule (fragrance fabrics) and fabrics that change color with temperature. An example of the color-changing fabrics is Toray's Sway brand of skiwear. This is PU-coated nylon that contains microcapsules containing heat-activated dye. A ski suit can be made to change color from bright red at the slopes outside and white indoors by the side of a fire. The color-changing fabric is of great interest because it has potential application for camouflage. The dyes used in these fabrics are thermochromic in nature which change color with temperature.

Numerous inorganic and organic compounds show thermochromism, and the subject has been reviewed [9]. Inorganic compounds show both reversible and irreversible thermochromism due to phase change or due to change in ligand geometry of the metal complexes. Temperature-indicating paints showing irreversible thermochromism has been used for a long time as a warning of hot spots and as a record of heat history in the electrical and chemical industries. The inorganic compounds have not found favor in textile applications as the color change generally occurs in solution or at high temperatures. The ideal thermochromic system for apparel application should show reversible thermochromism between ambient and body temperature. Reversible thermochromism in the solid state is exhibited by many organic compounds. These compounds undergo stereoisomerism, molecular rearrangements, or are liquid crystals. Among the liquid crystals, the most important systems are cholesteric mesophase. These have limited use in textiles due to high cost, color restricted to deep shades, low moisture resistance, and low color density. Sterically hindered ethylene compounds such as bianthrone and dixanthylene also show thermochromism. These compounds are characterized by at least one ethylene group, a number of aromatic rings, and a hetero atom, usually N or O. The ethylenic bond provides a route for extension of conjugation and places restrictions on possible molecular orientation. As the temperature is increased, the molecule changes to a different stereoisomer, which is colored. These compounds show transition above their melting point ~150°C, thus are not suitable for textile applications.

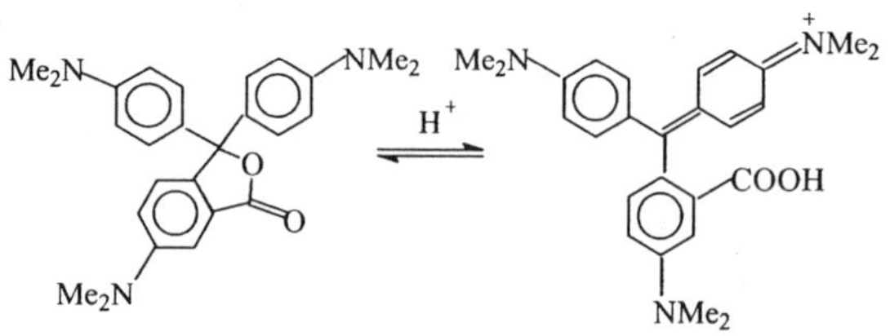

Figure 8.1 Tautomers of crystal violet lactone.

Certain dyes undergo keto-enol type of tautomerism. Such tautomeric rearrangements can lead to an increase in the conjugation and formation of a new chromophore, leading to color development. Such rearrangements can be induced by a change of temperature, pH, or polarity of the solvent, resulting in thermochromism. These dyes are extensively used for textile applications. The most common types are fluorans, crystal violet lactones, and spiro pyrans. All of these dyes undergo ring opening rearrangements [10]. The equilibrium of crystal violet lactone is shown in Figure 8.1.

Reversible thermochromism composition is made from such dyes, along with a color developer containing acidic protons capable of proton transfer or strong H bonding to the dye molecule. The dye and the developer are dissolved in a nonvolatile solvent, and the ternary composition is encapsulated. On heating, the organic solvent melts, resulting in color change. Some commonly used developers are bisphenol A, bisphenol B, 1,2,3-triazoles, thioureas, etc. Various compounds have been used as solvents, but the most common are aliphatic alcohols like stearyl alcohol. A puzzling feature of this system is that they are colored at low temperature and turn colorless at high temperature. Different reasons have been attributed to explain the same [10].

It has already been mentioned that the color development of the ternary system depends on the melting point of the solvent used. In order to ensure a homogeneous mixture throughout the color development stage, it is necessary to keep it in a closed system by microencapsulation. In a microcapsule, which is a small solid particle of $1-1000$ μm size, there is a core containing the thermochromic system and a coating or shell of a polymeric material. Two processes are prevalent in the literature for microencapsulation of thermochromic dye systems. They are complex coacervation and interfacial/in situ polymerization.

In complex coacervation, two polyelectrolytes of opposite charges are used, such as gelatin and gum arabic [11]. At a pH of <4.7, the gelatin is cationic, and gum arabic is anionic. The core material is initially dispersed in the gelatin solution. To this dispersion, a solution of gum arabic is added, and the pH is adjusted to ~ 4.0. This causes a liquid complex coacervate (droplets) of gelatin-gum arabic and water to form, which surrounds the core to form embryo capsules. The system is then cooled to gel the shell. The shell is then cross-linked with glutaraldehyde and dried to form a free-flowing powder of microcapsules.

The shell formation in interfacial polymerization occurs due to polycondensation at the surface of the core material. Core material and one of the reactants, like multifunctional acid chloride or isocyanate, are mixed together to form a water-immiscible mixture. This mixture is dispersed in water with the aid of an emulsifier. The other reactant, e.g., multifunctional amine or alcohol, is added to the aqueous phase. Interfacial polymerization occurs at the surface of the core, forming the shell of the microcapsule. In situ polymerization is the technique adopted for forming shells of aminoplasts. The solid core material is first dispersed in water that contains urea, melamine, or water-soluble urea-formaldehyde condensate. An anionic polymer is added to enhance aminoplast shell formation. Formalin is then added, and the pH is adjusted to 2–4.5. On heating (40–60°C), shell formation occurs.

The microcapsules are used as conventional pigments and are coated on fiber or fabric with the aid of polymeric binders.

Shibahashi et al. have described the above technology in coating a variety of fibers [12]. The fibers have been converted into yarns, nonwoven fabrics, and knitted and woven fabrics of various constructions. All of these show thermochromic effect. By proper selection of the dye system, they have been able to obtain thermochromic effects at temperatures ranging from −30 to 100°C. For uniform color change, proper pigment particle size has been selected depending upon the density and the denier of the fiber. The proportion of dye, developer, and solvent are critical for optimum results. Besides, the add on has to be carefully chosen to keep a balance between clear color change and the texture of the textile material. In a typical formulation, a thermochromic composition consisting of 1 part by weight of crystal violet lactone, 3 parts of benzoyl-4-hydroxy benzoate, and 25 parts of stearyl alcohol, was encapsulated by coacervation in gelatin-gum arabic. The microcapsules were coated on the fibers by dipping in a polyurethane emulsion. The resulting fiber exhibited reversible thermochromism, turning blue above 53°C and becoming colorless below that temperature. By coating the system on a dyed fabric, they have been able to achieve change of color from one colored state to the other. Different binders have been used for coating. They include low melting point thermoplastics, natural and synthetic resins, and emulsions. Coating has been done mainly by dipping or spraying. The applications include apparel, toys, artificial flowers, etc. Thermochromic patterns have been obtained on fabric using thermochromic and uncoated fibers.

8.3 TEMPERATURE ADAPTABLE FABRICS

Textile materials are being increasingly used for architectural purposes. However, these lack the thermal insulation of the conventional building materials. The thermal insulation can be increased by a double-shelled construction or by adding a layer of foam to the textile material. A new way to improve the thermal insulation can be the application of phase change materials (PCM). When a substrate containing PCM is heated, by solar radiation, the increase in temperature of the substrate is interrupted at the melting point of the phase change material, due to absorption as latent heat. The temperature will rise only when all the solid has melted. Conversely, during the cooling process at low ambient, the drop in temperature is interrupted at the solidification temperature. The heat flux through a material containing PCM is thus delayed in both heating as well as cooling, during the process of phase change. This thermal insulation effect is dependent on temperature and time; and being temporary in nature, it can be termed dynamic thermal insulation.

Vigo and Frost [13,14] have incorporated polyethylene glycol of different molecular weights as PCM in hollow fibers resulting in a 2–2.5 times increase in heat content compared to the untreated material. Similar results were obtained by treating textile materials with aqueous solution of the PCM by pad dry method. The main drawback of the process is that the PCMs are water soluble.

Considerable improvement in technique has been done by Pause [15], who has used hydrophobic higher hydrocarbons like dodecane, octadecane, etc., as PCMs. These compounds were encapsulated to form microcapsules of 1–60 μm size. The microencapsulated PCM was applied by a thin layer of lacquer on PVC-coated polyester with foam backing. In order to meet the requirements of widely varying ambient of winter and summer months, two PCMs having different transition temperatures were used. A procedure was devised to measure the dynamic thermal insulation properties. A comparative study of coated fabric containing 40 g/m^2 micro-PCM showed a fivefold increase in thermal insulation. The technology has great potential and further development work is in progress.

8.4 CAMOUFLAGE NETS

Camouflage nets are meant to conceal military equipments and objects from detection and attack by an enemy. It has a great effect also on the morale of the fighting forces. Historically, camouflage nets were first used during World War I. The earlier nets were made of hemp/cotton twine, garnished with jute strip scrims, and dyed/coated with green and brown colors. These nets blended in with the surroundings and prevented detection by naked eye or by binoculars. These nets had serious limitations, such as poor camouflage properties, premature fading of colors, susceptibility to fungus growth, short life, and high water absorption, resulting in a great increase in weight of the net when wet.

With the rapid developments of sophisticated surveillance systems such as active and passive infrared sensors, infrared line scanners (IRLS), forward looking airborne radar (FLAR), side looking airborne radar (SLAR), millimeter wave radar, etc., it became imperative to develop camouflage nets that would protect the objects from detection by various sensors. The requirements of camouflage net can be summarized as follows:

- small volume and light weight
- strong and durable
- waterproof and fire resistant
- colors and patterns similar to the surroundings
- easy to deploy, transport, and handle
- conceals objects against detection by sensors

To meet these requirements, modern camouflage nets are wholly made of synthetic textiles. The nets available offer different levels of protection, viz, against visual and near infrared; visual, near infrared and microwave; visual, near infrared, thermal infrared and microwave; and ultraviolet for snow terrains.

8.4.1 VISUAL AND NEAR IR NETS

These nets camouflage the objects in the visual (400–700 nm) and near IR (700–1200 nm): regions of the electromagnetic spectrum. That is, they offer protection against visual detection and against NIR sensors, such as night vision devices and image intensifiers. In order to blend the objects with the surroundings, the reflectance of the nets in the visible and near IR regions (400–1200 nm) should match that of the surroundings. To achieve these characteristics in the net, it is necessary to know the reflectance pattern of the objects constituting the surroundings in this spectral region. A typical spectrum of green vegetation containing chlorophyll is given in Figure 8.2, showing an IR reflectance value of about 50%. From such spectra, the IR reflectance of other inanimate objects of the surrounding area are obtained (see Table 8.4) [16,17].

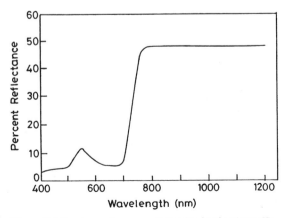

Figure 8.2 Spectrum of green vegetation in visual and near IR.

TABLE 8.4. IR Reflectance Values of Some Common Objects.

Constituents	Reflectance %
1. Green vegetation	50–70
2. Concrete	40–50
3. Damp soil	10–15
4. Dry soil	15–20
5. Sand	30–40
6. Building bricks	30–40
7. Galvanized iron	15–20

Two types of terrains are generally considered for camouflage, green vegetation and desert region. The color and IR reflectance of the net should match those of the terrain. Different nets are available for different terrains. A single net with reversible color pattern is also available.

The camouflage net consists essentially of two components, a netting forming the base and a garnishing material, usually coated fabric, that is fixed to the netting with clips. The netting is a square mesh of nylon twine, mesh size varying from 50–80 mm with soft vinyl coating or a flame retardant treatment. The garnishing material is usually PVC-coated nylon fabric incised in a suitable pattern. A lightweight fabric is taken as the base fabric ($\sim$70 g/m^2) that enhances the strength of the garnishing material. Unreinforced PVC films are also used for garnishing. The color scheme of the garnishing material depends on the terrain of deployment [18]. The colors for green belt are olive green, deep brunswick green, and dark brown. For the desert region, the colors are light stone, dark stone, and dark brown.

For effective camouflage in visual and near IR regions, the garnishing material should not only be of the desired color but should also have IR reflectance similar to that of the surroundings. For this purpose, pigment composition for PVC is first selected to obtain a visible match of the color. The level of IR reflectance is adjusted to the required value by introducing in the formulation normally a small percentage of high reflecting (e.g., TiO$_2$) or absorbing (e.g., carbon black) pigments. The nets are available in unit sizes of 1–1.7 m, which can be attached together to form a net of required size and shape. The incision in the garnishing replicates the shadows that occur in nature and breaks up the outline of an essentially rectangular or tubular nature of a vehicle or weapon system. The weight of the nets varies from 200–350 g/m^2.

8.4.2 RADAR CAMOUFLAGE NET

The most important component of a radar scattering net is the base fabric of the garnishing material. It consists of a specially designed nylon fabric in which is incorporated metallized yarns or aluminized polyester threads at regular intervals, both in warp and weft directions. The metallized threads form a regularly spaced grid structure in the fabric. The fabric is then coated with PVC compound containing pigments to meet the visual and near IR reflectance requirements. The coated fabric is incised in a definite pattern. By properly designing the grid structure of the metallized thread and with proper incision of the garnishing material, it is possible to scatter the incident microwave radiation from the radar, in a manner similar to that of the surroundings, leading to concealment of the object from radar. Nets made by Barracuda of Sweden have an attenuation of 10 dB against 3 cm radar (X band). It is claimed that the echo obtained from the object plus the net corresponds approximately to that of the surroundings [19].

8.4.3 SNOW CAMOUFLAGE

Snow has high reflectivity in UV region ~90%. A standard white shade, on the other hand, has a poor UV reflectance, ~10%. Thus, a military object when covered by standard white shade fabric in snow terrains, shows up as a black patch in a white background when viewed by reconnaisance devices using a UV filter at 350 nm. Camouflage nets for snow regions have a garnishing material of white PVC-coated nylon fabric having a UV reflectance of 75% minimum at 350 nm providing camouflage in both visual and UV regions.

8.5 METAL AND CONDUCTING POLYMER-COATED FABRICS

The incorporation of metal into textiles dates back to the Roman era, when they were mainly used for decorative purposes. The tinsel yarns used to add glitter to fabrics were made by flattening thin wire or sheets of noble metal like gold or silver. By the 1930s, aluminium foil strips coated on both sides by cellulose acetate-butyrate, to prevent them from tarnishing, were used. The yarn could be colored by anodizing. All of these yarns had poor compatibility with the more flexible and extensible textile yarns [20]. After the development of vapor-deposited aluminized polyester in the 1960s, 1 mm wide strips of these films were used as yarns, with much improved flexibility.

With the advancement of technology, metal/conductive textiles found extensive functional applications. These materials have high electrical conductivity and radar reflecting property, yet are lightweight and flexible. Various methods have been developed to coat fibers and textile materials by metals, and these are as follows [21]:

- coating metal powder with binders
- vacuum deposition

- sputter coating
- electroless coating

8.5.1 METHODS OF METAL COATING

a. Metal coating with a binder: the process is similar to conventional polymer coating. High leafing aluminium pastes (65–70%) are incorporated into a polymeric carrier, like synthetic rubber, PVC, polyurethanes, silicones, acrylic emulsions, etc., and spread coated on the fabric. The coating method may be conventional knife or roller coating. The adhesion, flex, and chemical resistance of the coated fabric depend on the type of polymer used, but they are not highly reflective.

b. Vacuum deposition: in this process, the substrate to be coated is placed in a chamber over a set of crucibles containing the metal to be coated in the form of a powder/wire. The chamber containing the whole assembly is evacuated to 0.5–1 torr. The crucible is heated by resistance heating to melt the metal. The temperature of heating is so adjusted that the vapor pressure of the metal exceeds that of the chamber pressure, so that substantial evaporation of the metal takes place. The temperature required for aluminium is about 1200°C. The roll of web to be coated is passed over a cooled drum placed over the crucibles. The metal atoms coming out of the molten metal hit the surface of the web to be coated and condense in the form of solid metal as it passes over the crucible. The production speed is quite high, ranging from 150–500 m/min. The items to be coated should be pretreated for proper adhesion of the metal. Continuous metal film coatings can be formed on just about any surface, film, fiber, or fabric with thicknesses ranging from micron to millimeter. Several metals can be vacuum evaporated, most common being aluminium, copper, silver, and gold. Difficulty arises in the case of metals, which sublime rather than melt and boil [21,22].

c. Sputter coating: the equipment consists of a vacuum chamber containing an inert gas, usually argon, at 10^{-3} to 10^{-1} torr (Figure 8.3). The chamber is equipped with a cathode (target), which is the source of the coating material, and an anode, which acts as a substrate holder. Application of an electrical potential of the order of 1000 Vdc, between the two electrodes, produces a glow discharge. A flow of current occurs due to movement of electrons from cathode to anode. The electrons ionize the argon gas. The argon ions are accelerated toward the cathode at a high speed due to high electric potential. The bombardment of the energetic ion on the target results in a transfer of momentum. If the kinetic energy of the striking ion is higher than the binding energy of the surface atoms of the material of the target, atoms are dislodged or sputtered from its surface by a cascade of collisions. Typically, the threshold kinetic energy of the ions should be between 10–30 ev for

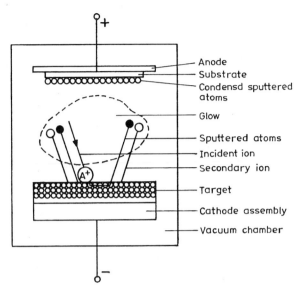

Figure 8.3 Sputtered coating process.

sputtering from the surface. Some of the ions striking the target surface generate secondary electrons. These secondary electrons produce additional ions, and the discharge is sustained. Considerable heat is generated during the sputtering process, and it is necessary to cool the target. The sputtered atoms and ions condense on the substrate to form a thin film of coating [23,24]. The relative rates of deposition depend on sputter yield, which is the number of atoms ejected per incident ion. The sputtering yield varies with the target material and increases with the energy of the incident ion. The method is applicable to a wide range of materials and gives more uniform coating with better adhesion than simple vapor deposition. The process is, however, more expensive, and the rate of deposition is lower ($\sim$30 m/min).

d. Electroless plating: it is a process to deposit metal film on a surface, without the use of electrical energy. Unlike electroplating where externally supplied electrons act as reducing agent, in electroless plating, metallic coatings are formed as a result of chemical reaction between a reducing agent and metal ions present in solution. In order to localize the metal deposition on a particular surface, rather than in the bulk of the solution, it is necessary that the surface should act as a catalyst. The activation energy of the catalytic route is lower than the homogeneous reaction in solution. If the deposited metal acts as a catalyst, autocatalysis occurs, and a smooth deposition is obtained [25,26]. Such an autocatalytic process is the basis of electroless coatings. Compared to electroplating, electroless coating has the following advantages:

(1) Nonconducting materials can be metallized.

(2) The coating is uniform.

(3) The process is simple and does not require electrical energy.

Electroless coating is, however, more expensive.

For successful deposition of coatings, only autocatalytic reduction reactions can be used. As such, the number of metals that can be coated are not many. Some of the common reducing agents are sodium hypophosphite, formaldehyde, hydrazine, and organoboron compounds. Each combination of metal and reducing agent requires a specfic pH range and bath formulation. The coating thickness varies between 0.01 μm to 1 mm.

A typical plating solution consists of

a. Metal salt

b. Reducing agent

c. Complexing agents, required in alkaline pH and also to enhance the autocatalytic process

d. Buffers

e. Stabilizers, which retard the reaction in the bulk and promote autocatalytic process

Some important metal coatings are discussed below.

a. Copper: the most suitable reducing agent is formaldehyde. The autocatalytic reaction proceeds in alkaline pH (11–14). The commonly used complexing agents are EDTA, tartarate, etc. The overall reaction is given by

$$Cu^{+2} + 2HCHO + 4OH^- \rightarrow Cu + 2HCOO^- + H_2 + 2H_2O$$

b. Nickel: sodium hypophosphite is the most popular reducing agent for nickel. The autocatalytic reaction occurs in both acidic and alkaline pH. Sodium citrate is used as buffer and complexing agent. The reaction is given as

$$Ni^{+2} + 2H_2PO_2^- + 2H_2O \rightarrow Ni + 2H_2PO_3^- + H_2 + 2H^+$$

The coating obtained by sodium phosphite also contains phosphorus (2–15%).

c. Silver: the plating solution consists of ammoniacal silver nitrate with formaldehyde, hydrazine, and glucose as reducing agents. Because the autocatalytic activity of silver is low, thick deposits cannot be obtained.

Nonconducting materials like polymers are given an etching treatment by chromic acid, followed by a catalytic treatment using stannous chloride or palladium chloride solution. Electroless plating of textiles is being adopted for different functional applications. The details of which are mostly covered by patents.

8.5.2 USES OF METAL-COATED FABRICS

Metallized fabrics and fibers find diverse applications, many of them in high-tech areas. Some of the important uses are described below.

8.5.2.1 Protective Clothing

The heat reflecting property of the metallized fabrics is used for protection against intense radiant heat for short duration [20,27,28]. Such suits are required by firemen during firefighting and workers in the steel industry, for protection against blast furnace radiaton and molten metal splash. The development of these suits has been stimulated by the widespread withdrawal of asbestos as a heat-resistant material. Three types of suits are in use by firemen that offer different levels of protection given by the thermal protection index (TPI). TPI radiation and flames are defined as the time in seconds for the temperature of the back surface of the clothing assembly to rise by 25°C above the ambient when exposed to a standard radiant heat source of 20 kW/M^2 at a distance of 200 mm or exposed to a standard heat source of burning hexane (BS 3791). All of these suits contain a heat-reflective fabric, which consists of a polyester film vapor deposited by aluminium to a thickness of about 200 Å on both sides and is laminated to glass fabric by a high temperature adhesive. The smooth surface of the polyester provides a high level of reflectivity.

a. Approach suit: this suit is meant for close approach to fires and protection against radiant heat only. It is made of different layers, the outermost being the heat-reflective fabric, followed by a neoprene-coated fabric as a moisture barrier, and preferably an inner layer of flame retardant cotton fabric, in contact with the body. It has a TPI of 50 against radiation (BS 3791).

b. Proximity suit: these suits are for operating in proximity to flame and offer protection from radiant heat and occasional flame lick. The suit consists of at least three layers: the outer shell made of heat-reflective fabric, a moisture barrier, and a thermal barrier (NFPA 1976). The insulation layer may contain Kevlar® or carbon fiber fleece. The TPI is 80 for radiation and 18 for flame.

c. Entry suit: the suit permits firemen to enter the flame for a short period for rescue operation. The suit contains several layers of heat-reflective fabric and insulating layers, but the outermost layer should be noncombustible, like asbestos or special glass fabric with conductive coating. The TPI is the highest for these suits—300 for radiation and 100 for flames.

Kiln entry suits for workers also contain an aluminized polyester heat-reflective layer.

8.5.2.2 Radar Responsive Fabrics

Metallized fabrics are capable of reflecting electromagnetic radiation and can act as a radar target by giving strong echo. They have advantages over metals including being lightweight and easier to fabricate into different objects [29]. One of the main applications of these fabrics is for making lifesaving devices for locating persons marooned in high seas. These include caps for lost fishermen, life jackets for aircraft crew forced to drop into the sea during an emergency, and foldable radar fabric reflectors for life rafts. Besides, movements of meteorological balloons are tracked by providing targets of metallized fabrics on them. In defense application, a target banner is towed behind an aircraft at a distance for practicing surface-to-air firing by soldiers after locating the same on a radar. The fabric for the banner is made of monofilament yarn (polyethylene, nylon, viscose) containing metallized threads of duralumin or silver, in both warp and weft directions, in plain-weave construction. Target parachutes and target sleeves are also used for similar firing practice.

8.5.2.3 Static Electricity Control

Rubbing action between two nonconducting materials tends to generate a static electrical charge. Some typical examples are walking on carpet, flow of hydrocarbon gas through plastic pipe, and reciprocal motion between textiles. The charge buildup may suddenly release in the form of a spark. This may cause fire or explosion in contact with flammable substances. Static electricity may also cause damage to electronic circuitry. The conductivity of common textile fibers is of the order of 10^{-13} $(\text{ohm} \cdot \text{cm})^{-1}$. Increasing the conductivity to 10^{-3} to 10^{-10} $(\text{ohm} \cdot \text{cm})^{-1}$ range is usually adequate for dissipation of static charge [30]. Some examples of antistatic fabrics are staff apparel in electronic industry, filter panels, conveyor belt reinforcement, antistatic flooring, etc. A common method to reduce static charge buildup is to impart a hygroscopic finish to the material by phosphoric acid ester, quaternary ammonium compounds, etc. These compounds absorb moisture from the atmosphere, increasing the conductivity to $\sim 10^{-10}$ $(\text{ohm} \cdot \text{cm})^{-1}$ for static control. The drawbacks of these compounds are that their effectiveness depends on the humidity of the environment and that they are removed by washing [20,30]. Incorporation of conducting fiber is a sure way of dissipating static charge in textile. These fibers are carbon fibers, metal fibers, and metallized fibers. Carbon fibers impart black color to the textile, while metal fibers are difficult to mix with other fibers due to their brittleness and high density. Metallized fibers obtained by vacuum deposition or electroless coating have the advantage of being processed like any other textile fiber. Development of copper-, nickel- and silver-coated fibers by chemical method has been reported for antistatic application [30–32].

8.5.2.4 Electromagnetic Interference Shielding

Any electrical or electronic device, including household appliances, generates electromagnetic radiation causing interference. These, in turn, can be disturbed by other devices. Various other sources of interference are cosmic rays, lightning, and high voltage power cables. The trend toward faster, more powerful electronic equipment and denser circuitry has increased the possibility of electromagnetic and radio frequency interference (EMI and RFI). This development presents a challenge to scientists to control EMI emission as well as to shield sensitive electronics from EMI to meet strict international regulations. Metallized fabrics are emerging as a material of choice for EMI/RFI shielding of sensitive equipment, particularly in defense and in aerospace. The conducting fabric can be tailored to ready-to-use adhesive tapes, curtains, bags, etc.

In order to understand the shielding process, let us consider an electromagnetic wave impinging on a shielding screen. The incident wave will undergo reflection from the surface, absorption by the material of the screen, and secondary reflection. The attenuated wave is transmitted from the other side of the screen. The shielding efficiency is expressed in decibel (dB) and is given by:

$$SE_{dB} = 20 \log_{10} E_I / E_T \quad \text{(for electrical field)}$$

$$SE_{dB} = 20 \log_{10} H_I / H_T \quad \text{(for magnetic field)}$$

where E_I and H_I are the incident electrical and magnetic fields, and E_T and H_T are the transmitted fields. The shielding efficiency increases with the conductivity of the fabric and thickness of the coating.

Various metal-coated fibers and fabrics have been reported in the literature for EMI shielding. Texmet brand of fibers are obtained by deposition of copper and nickel on acrylic fiber by a chemical process. A coating thickness of 0.3 μm gives a conductivity of 10^3 (ohm $\cdot$ cm)$^{-1}$. The fiber is available in the form of crimped staple fiber or as continuous tows. Nonwoven fabrics of different conductivities have been made by incorporating Texmet fiber into other textile fibers, like polyester, polypropylene, and acrylic. Nonwoven with 70% Texmet loading showed very good microwave reflection and a transmission loss of over 65 dB in the 8–12 GHz range [30]. Temmerman [31] has described an electroless coating process known as Flectron (a brand name of Monsanto Chemical Co.). Metal coating of copper, nickel, silver, etc., singly or a combination of one metal with an overcoat of a second metal has been done by this process on a variety of substrates, including woven, knitted, nonwoven fabrics, chopped strands, and films. The metal content of these fabrics ranges from 14–24 g/m^2, with surface resistivity varying from 0.04–(0.43 ohms/sq (see AATCC test method 76-1995). A coating of 15 g/m^2 metal, Cu, Cu/Ni, or Sn/Cu on nylon nonwoven gave a shielding between 56–90 dB in the frequency range of 100 to 10,000 MHz. It was estimated that these metallized fabrics provide 95–97% of the shielding than would be provided by an equivalent mass of metal in foil form. The metal layers are well adhered on the substrate. Silver-coated nylon fiber and nylon fabric have also been developed by Statex system of coating for different applications [32], including EMI shielding. Shielding of over 40 dB has been obtained in 90 GHz to 500 MHz range using silver-coated nylon fabric.

Apart from the above specialized applications of metal-coated fibers/fabrics, they are also used for several other purposes. The passage of an electric current through metal-coated panels of fabric results in resistive heating. This property can be used for making heated garments, gloves, blankets, and as an IR tank decoy [33]. Some other uses are ironing board covers and pleated window shades for thermal protection. A comprehensive list of applications of these fabrics has been given by Smith [21].

8.5.3 CONDUCTING POLYMER COATINGS

Several people have been working on the development of conducting polymer-coated textiles. Major problems of these coatings are their poor environmental stability and difficulty in processing them from solution or melt. As such, the technology is still in the developmental stage. The polymers that have been tried are polypyrrole and polyaniline. Coating of polypyrrole on textiles has been done by Jolly et al. [34] in a one-step process, in an aqueous solution containing the monomer, $FeCl_3$ oxidant and napthalene sulphonic acid dopant, at a reaction temperature of 5–10°C. Polymerization occurs at the surface of the textile, and each fiber is coated with a homogeneous polypyrrole layer. The coated fabric has a surface resistivity of about 10 ohms/sq and has been tried as heating panels for buildings. Effect of aging on the conductivity has also been studied at different temperatures. Jin and Gong [35] have deposited polyaniline on polyester fiber and nylon fabric by aniline diffusion and oxidative polymerization, using HCl as a dopant. The process consists of immersing the textile in aniline and removing the absorbed aniline from the surface by treatment with hydrochloric acid. The specimens were then treated with aqueous aniline hydrochloride solution, followed by oxidation with ammonium persulphate, and washed. Aniline diffusion and use of HCl as dopant enhanced the conductivity and adhesion of the coating. Polyaniline and polypyrrole coatings have also been studied by Trivedi and Dhawan [36] and Gregory and coworkers [37].

8.6 REFERENCES

1. P. K. Ramchandran, and N. Raja, *Defence Science Journal,* 40, 1990, pp. 15–23.
2. T. P. Carroll, *Journal of Coated Fabrics,* vol. 24, 1995, pp. 312–327.
3. *Janes NBC Protective Equipments,* 7th Ed. , T. J. Gander, Ed. , 1994–1995, Jane's Informative Group Ltd., U. K., pp. 18–22.
4. S. N. Pandey, A. K. Sen, and V. S. Tripathi, *Man Made Textiles in India,* May, 1994, pp. 185–188.
5. F. A. P. Maggs, P. H. Schwabe, and J. H. William, *Nature,* 186, 1960, pp. 956–958.
6. K. Gurudutt, V. S. Tripathi, and A. K. Sen, *Defence Science Journal,* 47, 1997, pp. 239–240.
7. *Textile Horizons,* vol. 8, no. 1, 1988, p. 7.
8. *Textile Horizons,* vol. 8; no. 12, 1988, p. 45.
9. *Kirk Othmer Encyclopedia of Chemical Technology,* 3rd Ed., vol. 6, 1979, John Wiley and Sons, New York, pp. 129–142.
10. D. Aitken, S. M. Burkinshaw, J. Griffith, and A. D. Towers, *Review of Progress in Coloration,* vol. 26, 1996, pp. 1–8.
11. *Kirk Othmer Encyclopedia of Chemical Technology,* 4th Ed., vol. 16, 1995, John Wiley and Sons, New York, p. 630.
12. Y. Shibahashi, N. Nakasuji, T. Kataoka, H. Inagaki, T. Kito, M. Ozaki, N. Matunami, N. Ishimura, and K. Fujita, U.S. Patent 4,681,791, 1987.
13. T. L. Vigo, and C. M. Frost, *Journal of Coated Fabrics,* vol. 12, April, 1983, pp. 243–254.
14. T. L. Vigo and C. M. Frost, *Textile Research Journal,* Dec., 1985, pp. 737–743.
15. B. Pause, *Journal of Coated Fabrics,* vol. 25, July, 1995, pp. 59–67.
16. R. Indushekhar, A. Srivastava, and A. K. Sen, *Man Made Textiles in India,* Dec., 1996, pp. 449–453.
17. S. M. Burkinshaw, G. Hallas, and A. D. Towns, *Review of Progress in Coloration,* vol. 26, 1996, pp. 47–53.
18. J. Musgrove, *Indian Textile Journal,* July, 1991, pp. 24–33.
19. Technical literature of M/S Barracuda, Sweden.
20. J. E. Ford, *Textiles,* vol. 17, no. 3, 1988, pp. 58–62.
21. W. C. Smith, *Journal of Coated Fabrics,* vol. 17, April, 1988, pp. 242–253.
22. Metal coatings, R. D. Athey Jr. , in *Coating Technology Handbook,* D. Satas, Ed., Marcel Dekker, New York, 1991, p. 691.
23. Sputtered thin film coatings, B. Aufderheide, in *Coating Technology Handbook,* D. Satas, Ed., Marcel Dekker, New York, 1991, p. 217.
24. *Kirk Othmer Encyclopedia of Chemical Technology,* 3rd Ed., vol. 15, 1979, John Wiley and Sons, New York, p. 264.
25. Electroless plating, A. Vaskelis, in *Coating Technology Handbook,* D. Satas, Ed., Marcel Dekker, New York, 1991, p. 187.
26. *Kirk Othmer Encyclopedia of Chemical Technology,* 4th Ed., vol. 9, 1995, John Wiley and Sons, New York, pp. 198–218.
27. R. A. Holmberg, *Journal of Coated Fabrics,* vol. 18, July, 1988, pp. 64–70.
28. *Wellington Sears Handbook of Industrial Textiles,* S. Adanur, Ed., Technomic Publishing Co., Inc., Lancaster, PA, 1995.

29. J. K. Tyagi, *Indian Textile Journal,* Aug, 1986, pp. 76–80.

30. F. Marchini, *Journal of Coated Fabrics,* vol. 20, Jan., 1991, pp. 153–165.

31. L. Temmerman, *Journal of Coated Fabrics,* vol. 21, Jan., 1992, pp. 191–198.

32. K. Bertuleit, *Journal of Coated Fabrics,* vol. 20, Jan., 1991, pp. 211–215.

33. R. Orban, *Journal of Coated Fabrics,* vol. 18, April, 1989, pp. 246–254.

34. R. Jolly, C. Petrescu, J. C. Thiebelmont, J. C. Marechal and F. D. Menneteau, *Journal of Coated Fabrics,* vol. 23, Jan., 1994, pp. 228–236.

35. X. Jin, and K. Gong, *Journal of Coated Fabrics,* vol. 26, 1996, pp. 36–43.

36. D. C. Trivedi, and S. K. Dhawan, in *Proc. of Polymer Symp. 1991,* Pune, India, S. Sivaram, Ed., McGraw Hill, India, 1991.

37. R. V. Gregory, W. C. Kimbrell, and H. H. Kuhn, *Journal of Coated Fabrics,* vol. 20, Jan., 1991, pp. 167–175.

CHAPTER 9

Test Methods

THERE are a number of tests used to evaluate coated textiles. The basic principles and relevance of the tests typical to coated fabrics are discussed in this chapter. Tests that are common to uncoated textiles, such as roll characteristic (length, width, and mass), breaking strength, tear resistance, and bursting strength, have not been included. ASTM-D 751-79 prescribes that all tests are to be carried out after a lapse of at least 16 h between curing and testing. Specimens are cut in such a way that no specimen is nearer than one-tenth of the width of the fabrics. The conditioning of the specimens is done at specified conditions of temperature and humidity depending on the standard used, i.e., ASTM, BS, Indian Standard, DIN, etc.

9.1 COATING MASS PER UNIT AREA (BS 3424, IS 7016 PART 1)

Three test specimens of 2500 mm^2 areas in circular, rectangular, or square shapes, are cut, conditioned, and weighed. The result is expressed in g/m^2. The coating of these specimens is then removed by selecting a proper stripping solution for the particular nature of coating. The bulk of the coating is removed mechanically by wetting the specimens with stripping solvent. The specimens are then refluxed with the solvent, washed with acetone, dried, and weighed. The process is repeated until the difference is <1% between successive refluxing and washing. The stripping solvent for PVC is tetrahydrofuran or methyl ethyl ketone. For natural rubber on cotton, nitrobenzene/xylene is used. PU-coated fabrics are stripped by 2 N alcoholic KOH. The weight of the base fabric thus obtained is also expressed in g/m^2. The coating mass in g/m^2 is obtained by subtracting the weight of the base fabric from that of the coated fabric.

In a rubberized fabric, the rubber hydrocarbon (polymer) content can be determined either by indirect method or by direct method. In the indirect method, the nonrubber ingredients are estimated from the contents of acetone, chloroform, alcoholic potash extracts, and determination of fillers and sulfur contents.

203

The rubber content is obtained by difference. This method is applicable for all olefinic rubbers. Direct method covers the determination of specific rubber polymer in the product based on the estimation of an element or a functional group. Thus, natural rubber content can be estimated from the acetic acid generated on its oxidation by chromic acid. In a similar manner, estimation of nitrogen or chlorine by standard techniques permits calculation of nitrile rubber or neoprene content, respectively (IS 5915 and 6110).

9.2 DEGREE OF FUSION/CURING OF COATING (BS 3424, ASTM D 4005-81)

PVC is generally coated as a dispersion in a solvent by the spread-coating process. After coating, the layer is fused to form a uniform film. During fusion, phase inversion occurs. Proper fusion determines the durability of the coating. Three test specimens of 40×25 mm are cut from the roll of the fabric. The specimens are immersed for 15 min in acetone at 20°C (BS) or 30 min at 23°C (ASTM), and the coating is examined. If there is no cracking or disintegration of the coating, disregarding surface effect or removal of lacquer, the sample is considered to have passed.

For rubberized fabrics, the specimens are immersed in xylol for 2 h at 27°C, and the coating is examined. The sample is considered to have cured if there is no tackiness in the coating or no detachment from the base fabric (IS 9491).

9.3 BLOCKING (BS 3424, IS 7016 PART 9)

This test is to check the tackiness of the coating at elevated temperature, so that the coating is not damaged when stored in rolls. In the BS method, two specimens of 150×75 mm are cut from the roll, placed face to face covering each other completely, placed in an oven at 60°C with a 1.5 kg weight piece placed over it, covering half the area of the specimen pair, and kept for 15 min. The specimens are then taken out, and a 100 g weight is hooked on the free end of the lower strip. No blocking is reported if the upper strip can be separated from the bottom strip at 25 mm/s rate of pull without lifting the 100 g weight piece, and there is no visible damage to the surface of the specimen.

In the IS method (based on ISO/DIS 5978-86), six specimens each of 150×150 mm are cut. The specimens are piled in three pairs, back to back, back to face, and face to face. The three pairs are placed in such a way that 100 mm^2 pile is formed, leaving the rest of the area free. The pile is placed between two glass plates in an oven at 70°C. A 5 kg weight piece is placed over the pile assembly. The specimens are taken out after 3 h, cooled at ambient for 3 h, and examined. No blocking is reported if the specimens can be separated without any sign of adhesion. This method examines blocking between the coated and uncoated surfaces.

9.4 COATING ADHESION (BS 3424, ASTM D-751, IS 7016 PART 5)

This test is of importance because if the adhesion is inadequate, separation of the coating from the base fabric may occur. As per BS, test specimens of 75 × 200 mm with length perpendicular to the longitudinal axis are cut. If the coating is thick, i.e., where the strength of the coating is more than the adhesive bond between the coating and the fabric, the coating is manually stripped to about 50 mm, and the width of specimen is trimmed to 50 mm. The adhesion strength can be determined by either a dynamic method or a dead weight method. In the dynamic method, the separated plies of the specimens are clamped to the jaws of an autographic strength testing machine with constant rate of traverse. Coating is separated for about 100 mm by setting the traverse jaw in motion. The adhesion strength is obtained from the load required to separate the coating layer.

The dead weight test apparatus consists of two grips, the top fixed to a rigid support and the bottom free, capable of accepting dead loads of 200 g units. The separated plies of the specimen are attached to the two grips, and dead weight is gradually placed on the lower jaw until separation occurs within a specified rate (5 mm in 5 min). This load is recorded. If the thickness of the coating is thin, two specimens are bonded face to face by an adhesive system leaving 50 mm free. For vinyl coating, the adhesive may be a solution of PVC resin in tetrahydrofuran, and for PU coating, suitable PU adhesive is taken. At the adhesion line of the two specimens, one layer of fabric and both layers of coating are cut and manually stripped to a specified distance. The adhesion strength is determined as above by fixing one layer of base cloth in the fixed jaw and the composite layer of two coatings and base cloth in the movable jaw.

ASTM and IS procedures are similar except that they specify only a dynamic tensile testing machine for adhesion and that the shape of the peeled layer of coating of the specimen is in the form of a tapered tongue (Figure 9.1).

9.5 ACCELERATED AGING (BS 3424, IS 7016 PART 8 BASED ON ISO/R 1419-1970)

Two methods are prescribed for accelerated aging: oven method and oxygen pressure method. In oven aging, specimens are heated in an air oven at 70°C or 100°C, as required, for 168 h. After the exposure, the nature of the coating is observed for any sign of softening, stiffening, brittleness, or sticking. Determination of a physical property before and after aging permits the calculation

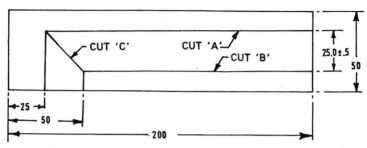

Figure 9.1 Dimensions and cutting line of test piece (IS 7016 Part V). Cut A extends from one end of the test piece to within 25 mm of the other end. Cut B extends to within 50 mm of that end. Cut C is a diagonal cut joining the ends of cut A and B (all dimensions in mm). (Adapted with permission from Bureau of Indian Standards.)

of percentage loss of the physical property on aging. In the oxygen pressure aging test, the specimens are subjected to elevated pressure and temperature by hanging them vertically in a pressure chamber of stainless steel at an oxygen pressure of 2000 kN/m^2 and 70°C for 24 h.

For vinyl-coated fabrics, estimation of loss of plasticizer is of great importance as it gives an idea of the durability of coating on weathering of these fabrics. ASTM-D1203 describes estimation of loss of volatiles from coated fabrics under defined conditions of time and temperature using active carbon as the immersion medium. In the direct contact method (A), three specimens are placed in a covered container with alternate layers of active carbon of specified particle size. The container is heated in an oven at 70°C for 24 h. The weight loss of the specimens is measured. Method B, known as the wire cage method, is similar to method A, except that the specimens are placed in the annular space of a cylindrical metal cage made of bronze gauze and are not in direct contact with active carbon. The wire cage method is similar to the BS method. In the IS method (IS 1259), the loss of mass as volatiles is estimated by exposing test specimens at 100°C for 24 h in an air oven.

For elastomer-coated fabrics, ASTM D3041 describes an ozone cracking test. The specimens are under strain by placing them around a mandrel and are exposed in a ozone test chamber containing an atmosphere of ozone and air (50 mPa partial pressure of ozone), and at a temperature of 40°C. The specimens are examined for cracks by magnifying glass after exposure for a specified duration of time.

A stringent accelerated aging test has been given for inflatable restraint (air bag for automobiles) fabric in ASTM D 5427-95. The fabric specimens are evaluated for selected physical properties after cyclic, heat, humidity, and or ozone aging. In cyclic aging, the specimens are aged in a specified cycle of temperature and humidity conditions. Typical aging cycle conditions are (a) −40°C, ambient RH for 3 h; (b) 22°C, 70% RH for 2 h; and (c) 107°C, ambient RH, for 3 h. The conditions for heat aging, humidity aging, and ozone aging

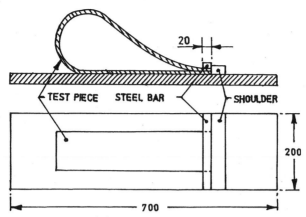

Figure 9.2 Apparatus for flexibility determination (IS 7016 Part 11) (all dimensions in mm). (Adapted with permission from Bureau of Indian Standards.)

are 105°C for 400 h; 80°C, 95% RH for 336 h; and 40°C, 65% RH, 100 pphm ozone concentration for 168 h, respectively.

9.6 FLEXIBILITY—FLAT LOOP METHOD (IS 7016 PART 11 BASED ON ISO 5979-1982)

This simple test measures the flexibility of rubber- or plastic-coated fabric. Rectangular strips of 600 mm × 100 mm are cut. A loop is formed from the strip and placed on a horizontal plane by superposing the two ends that are held in place under a steel bar (see Figure 9.2). The height of the loop is measured, which gives an idea of the flexibility of the fabric. The lower the loop height, the greater the flexibility, and vice versa.

9.7 DAMAGE DUE TO FLEXING (BS3424, IS 7016 PART 4, ASTM D 2097)

Accelerated flexing of coated fabrics gives useful information of the durability of the coating in actual use. One of the common methods described in the BS/IS standards for flex testing is the De Mattia method. The apparatus consists of pairs of flat grips. The grips of each pair are positioned vertically one over the other, and one grip is capable of reciprocating motion in a vertical plane. The traverse distance of the grips in open and closed positions is 57 mm. The rate of reciprocating motion of the grips is specified (300 cycles/min). Test specimens of 45 × 125 mm are cut with length in longitudinal and cross directions. Each test specimen is folded with coating outwards along lines 15 mm from each of the longer sides and to a width of 15 mm. The specimens are then clamped between the grips of the equipment and flexing is carried out for specified cycles (~100,000 cycles). The specimens after flexing are examined for number of cracks, their severity, and delamination.

ASTM D 2097 is meant for upholstery leather but has been adopted for vinyl upholstery fabric (ASTM D-3690). The testing is carried out in Newark-type flexing machine containing a pair of pistons. One piston of the pair is stationary, and the other is movable. The piston moves with a reciprocating motion in a horizontal plane at 500 rpm with a stroke of 32 mm. The closed position of the piston is adjusted to 15 times the thickness of the specimens. The sizes of test specimens are 76 × 114 mm and are clamped on the pair of pistons in a cylindrical shape. After a predetermined number of cycles, the fabric is visually examined for cracks.

9.8 ABRASION RESISTANCE (ASTM D-3389, BS 3424)

The abrasion resistance of a coated fabric is determined by abrading the coated surface of the fabric with an abrader. Measurement of mass loss after abrasion gives an idea about the abrasion resistance of the coating. In the ASTM method, a revolving double-headed platform (RPDH) abrader is used. Circular test specimens of 110 mm diameter are cut and placed with the coated side up on a specimen holder affixed on a rotating platform. The platform is rotated in a circular motion at 70 rpm. The abrasion is done by two abrasive wheels made of a specified material that are attached to the free end of a pair of pivoted arms. The abrasive wheels rest on the specimen in a manner that a vertical force is applied on the specimen by them. The force can be increased by the addition of weights. The abrasion occurs due to friction between the rotating specimen and the abrasive wheels. The loose abraded particles are removed from the specimen by vacuum cleaner. From the mass loss of the specimen for specified cycles of operation, the mass loss per revolution is estimated.

The BS specifies the Martindale abrasion tester for testing expanded PVC coating. The machine consists of a rectangular base plate on which are mounted four circular discs covered with specified silicon carbide paper. A top plate containing a four-specimen holder rests on the center of the abrading discs. Four circular test specimens are cut and placed on the specimen holder, with coated surface facing the abrading disc. A specified load is applied on the specimens by placing desired weights. The abrasion occurs due to the rotation of the plate holding the specimens, such that the specimen rubs against the abrading disc in a definite pattern. The pattern traced by the plate is similar to Lissajou's figures, i.e., it changes from a circle to gradually narrowing ellipses to a straight line followed by gradually widening ellipses to a circle. After a specified number of cycles, the exposure of the cellular layer of the coated fabric is noted.

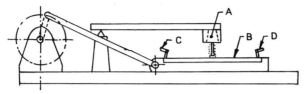

Figure 9.3 Apparatus for measuring colorfastness to rubbing (IS 1259): (A) abrader member, (B) glass plate, and (C, D) grips. (Adapted with permission from Bureau of Indian Standards.)

9.9 TEST FOR COLORFASTNESS TO DRY AND WET RUBBING (BS 3424, IS 1259)

The test essentially consists of rubbing the coated fabric specimens with white fabric with a specified load and number of cycles and examining the stain imparted, if any, on the white fabric from the test specimen. The test is also known as colorfastness to crocking and is carried out in a crockmeter. Test specimens of 230 × 50 mm are cut from the roll and mounted with coated side up on a flat glass surface. A circular piece of bleached white fabric of 25 mm diameter is affixed to a circular brass abrading peg of 16 mm diameter. The abrading peg is fixed to a pivot by an arm. The peg is imparted a reciprocating motion in a straight line parallel to the surface of the test specimen, either manually or mechanically, with a stroke of 100 mm at a rate of 15 cycles/min and is loaded in a manner to exert 0.5 kgf on the test piece. In dry rubbing, the staining of the cotton fabric is examined after ten abrading cycles and compared with the gray scale. In wet rubbing, the cotton fabric is wetted by diluted soap-soda solution prior to rubbing. Apparatus specified in the IS standard is shown in Figure 9.3.

9.10 LOW TEMPERATURE BEND TEST (ASTM D 2136, IS 7016 PART 10, BASED ON ISO 4675-90)

Coated fabrics are used in many applications requiring low temperature flexing. This method is meant for evaluation of the ability of rubber- and plastic-coated fabrics to resist the effect of low temperature, when subjected to bending.

Three test pieces of 25 × 100 mm are cut from the rolls. Each specimen after conditioning is placed between a pair of glass plates, to prevent curling, and then placed into a low temperature cabinet maintained at the specified temperature. A bending jig for bending the sample after exposure is also placed in the cabinet. The bending jig consists of two rectangular aluminium blocks. The blocks are connected by a hinge and aligned in a straight line. The two blocks are mounted in a frame at an angle of 60° from the horizontal (Figure 9.4). To the top block is attached a 250 g weight. A release mechanism folds the hinge at 180° so that the upper plate with weight falls free for bending the specimen. After the

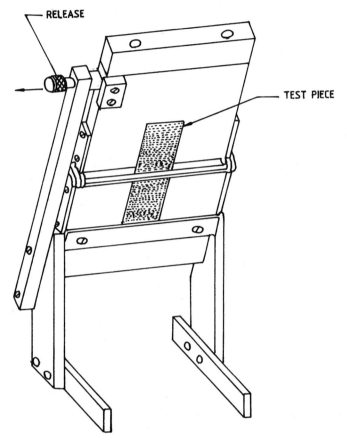

RELEASE

TEST PIECE

Figure 9.4 Bending jig for low temperature bend test. (Adapted with permission from Bureau of Indian Standards.)

end of the exposure period (4 h), sample specimen is placed in the jig and bent within the cabinet itself by the release mechanism. The samples are then taken out, folded at 180°, and examined for cracks and their severity.

9.11 LOW TEMPERATURE IMPACT TEST (BS 3424, ASTM D-2137, IS 7016 PART 14)

This is another test to determine the applicability of rubber- and plastic-coated fabrics at low temperatures. By this method, the lowest temperature at which the fabrics do not exhibit cracks in the coating when subjected to specified impact conditions is measured. In the ASTM method, test specimens of 6.4 × 40 mm are die punched, conditioned, and one end is clamped in a specimen clamp designed to hold the specimen as a cantilever beam, such that

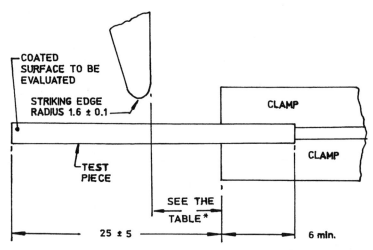

Figure 9.5 Test piece holder and striker in low temperature impact test. *Refers to Table 1 of IS 7016 pt. 14 specifying clearance of striking arm and test piece clamps for test pieces of different thicknesses (all dimensions in mm). (Adapted with permission from Bureau of Indian Standards.)

the length extending from the clamp is 25 mm. A solenoid-activated striker is positioned on top of the specimen to impart an impact on it. The radius of the striking arm, the position of strike at the specimen from the clamp, the traverse distance of the striker, and the speed of traverse are specified. The test piece holder and striker specified in IS are shown in Figure 9.5. The cold crack temperature varies with the rate of folding, and as such, it is essential to fix the impact velocity at 2 m/s. The test assembly (specimen, clamp, and striker) is immersed in a low temperature bath of methanol, silicone, or other suitable heat transfer fluid, or in a refrigerated cold cabinet. After temperature equilibration, impact is applied on the specimen through the striker and cracks in the coating examined by taking out the specimen. The temperature of the bath is lowered by intervals at 10°C until the specimen fails. The temperature of the bath is then raised by 1°C intervals until the specimen passes. The temperature 1°C below this point is the cold crack temperature.

In the BS method, a folded test piece in the form of a loop is placed on an anvil and immersed in a low temperature bath by a holder. Impact is provided by a spring-actuated hammer of specified weight. The impact velocity is maintained at 2 m/s. Cold crack temperature is determined in a similar manner.

9.12 CONE TEST (IS 7941)

This is a test for the waterproofness of the fabric that is locally stressed. A circular/rectangular specimen of fabric is folded twice and then opened to obtain a cone with the coated side inward. At the tip of the cone, the material is folded so sharp that the coating is heavily stressed. The cone is then put in a wire cone and, in turn, placed into a glass funnel and filled with a specified quantity of water. There should not be any penetration of liquid, as well as no wetting of the outer surface of the specimen cone after 18 h.

9.13 RESISTANCE TO WATER PENETRATION (ASTM D-751, BS 3424, IS 7016 PART 7)

This test is of great relevance for coated fabrics because it evaluates the continuity of coating film and its resistance to water penetration, which is an important property of certain fabrics, particularly those used as rainwear, covers, and inflatables. ASTM describes two methods. Method A uses a Mullen-type hydrostatic tester, and in method B, pressure is applied by a rising column of water.

In a Mullen-type hydrostatic tester, a test specimen is clamped between two circular clamps having an aperture of 31.2 mm diameter. Hydraulic pressure is applied to the underside of the specimen by means of a piston forcing water into a pressure chamber. The pressure is measured by a Bourdon's gauge. During testing, steadily increasing pressure is applied on the specimen, and the pressure is noted when water first appears through the specimen. Alternately, a specified pressure is applied on the specimen for 5 min and appearance of water through the fabric is noted. A specimen is considered to have passed if there is no leakage of water at that pressure. Method A is not applicable for fabrics having water resistance less than 35 kPa pressure.

In method B, the test specimen is mounted on a ring with a conical bottom with the coated surface in contact with water. The specimen is clamped by placing a dome-shaped movable water chamber at the top. The water chamber has a water inlet and a vent. A water leveler consisting of a water inlet, a water outlet, and an overflow pipe is attached to the inlet of the chamber and is the means for setting the head of water. The overflow pipe regulates the level of water. During test, the head of water is increased by raising the water leveler by a motorized system, at the rate of 1 cm/s. The pressure at which the first drop appears through the underside of the specimen is noted. In an alternative procedure, the head of water is kept steady for a specified period, and leakage, if any, is noted. The low pressure method of BS, IS, and the hydrostatic pressure test of AATCC 127 are similar.

9.14 AIR PERMEABILITY (BS 3424)

The air permeability of coated fabric is low. This method is used primarily for porous discontinuous coating of breathable fabric. Air permeability is defined as the volume in milliliters of air that passes through the fabric per second, per cm^2, at a pressure of 1 cm head of water. The apparatus described in the standard consists of a specimen holder, in which the specimen is clamped between two flanges having an orifice of 25 mm diameter (a test area of 5.07 cm^2). Air is sucked through the test specimen by means of a vacuum pump. The rate of flow is adjusted by a series valve and a bypass valve, setting the pressure drop of 1 cm water head, across the fabric which is indicated by a dial gauge. The rate of flow of air is measured by an appropriate rotameter when steady pressure drop of 1 cm head of water is achieved. The apparatus is based on the Shirley Institute, U.K., air permeability apparatus.

9.15 WATER VAPOR PERMEABILITY (ASTM E-96-80)

Water vapor permeability is an important parameter of breathable fabric, as it gives an idea about the comfort property of the fabric. The ASTM method is, however, a general method for all materials, like paper, plastic, wood, etc., and not specifically for textiles. Two methods are described, viz., the desiccant method and the water method. The details of the two methods are described below.

9.15.1 DESICCANT METHOD

For the test, a shallow test dish is taken. The weight and size of the dish should be such as that can be weighed in an analytical balance, having a mouth of at least 3000 mm^2 area. The dish should have a ledge around the mouth to fix the test specimen. A layer ($>$12 mm) of anhydrous calcium chloride of specified mesh size is filled in the dish within 6 mm of the specimen. The mouth of the test dish is covered with test specimen, and the edges are sealed by molten wax. The whole assembly is weighed and placed in an air circulated, temperature and humidity controlled test chamber. The temperature of the chamber can be maintained at a selected value, but a temperature of 32°C is recommended. The humidity of the chamber is maintained at 50% or 90% RH depending on the environment desired. The weight gain of the test assembly is measured periodically, and at least 10 data points are taken during the duration of the test. A plot of weight against elapsed time is drawn. The slope of the straight line plot gives G/t (where G is weight change in gms and t is time in h). The water vapor transmission is given by WVT (g/h/m^2) $= (G/t)A$ (A is the area of the exposed specimen in m^2).

9.15.2 WATER METHOD

In this procedure, the test dish is filled with distilled water to a depth of 3–5 mm, with air gap of about 20 mm from the specimen. The procedure is the same as that for the desiccant method. The weight loss is measured periodically. If the barrier material is expected to be in contact with water in service, the dish is inverted during the test.

The BS 7209 method for estimation of water vapor permeability index % (*I*), for breathable fabric, is similar to the water method of ASTM. However, in this specification, the test specimens are tested along with a specified reference fabric, for water vapor permeability, and from the ratio of their water vapor permeability, *I* is calculated. The reference fabric is made of monofilament high tenacity woven polyester yarn of 32 μm diameter having an open area of 12.5%. As per the procedure, test specimens/reference fabric are sealed over the open mouth of a test dish, with cover ring of specified dimensions, containing distilled water. The quantity of water is adjusted to maintain a still air of 10 ± 1 mm between the underside of the specimen and the surface of the water. A sample support placed on the mouth of the dish prevents sagging of the specimen and the resultant change in the depth of the still air layer. The whole test assembly is placed on a rotating turntable. The turntable with the test assemblies are, in turn, placed in an environmental chamber, maintained at 65% RH, and 20°C temperature. The turntable is rotated at a slow specified rate to avoid formation of a still air layer above the dish, care being taken that the depth of still air is not altered inside the dish due to rotation. The test assemblies are weighed after a period of $\sim$1 h to permit equilibration of water vapor gradient in each assembly. Dishes are then placed back on the turntable in the chamber, and the test is continued for a period of $\sim$16 h, after which the assemblies are reweighed. From the loss of mass of the assemblies between the two weighings, the index *I* is calculated.

$$\text{Water vapor permeability in g/m}^2/\text{day} = 24M/At$$

where $M =$ loss of mass in g of assembly in time t (h), and $A =$ area of the test fabric in m^2 exposed.

$$I = \{(wvp)_f/(wvp)_r\} \times 100$$

where $(wvp)_f$ and $(wvp)_r$ are the mean permeability of test specimens and that of reference fabric, respectively.

9.16 RESISTANCE TO PERMEATION BY HAZARDOUS LIQUID CHEMICALS (ASTM F 739-96)

This test determines the resistance to permeation of a hazardous liquid through protective clothing material under continuous contact. The apparatus consists of a glass cell with two compartments, with the test specimen placed in between. One compartment is filled with the hazardous liquid, while the other contains a collecting fluid to absorb the permeated material. The test specimen acts as a barrier between the challenge chemical and the collecting medium. The collecting fluid may be a liquid or a gas in which the hazardous liquid is freely soluble. The chemical permeating through the specimen dissolves in the collecting fluid and is analyzed continuously or discretely by a suitable analytical technique. Some common techniques are UV, IR, spectrophotometry, gas liquid chromatography, and colorimetry. The barrier property of the material of the specimen is evaluated by measuring the breakthrough time (BTT) as well as permeation rate. The BTT is the time elapsed in seconds between the initial contact of the chemical with the outside surface of the specimen and the time at which the chemical can be detected at the inside surface by the analytical tool.

9.17 RESISTANCE TO PENETRATION/PERMEATION OF CHEMICAL WARFARE AGENTS [1]

These tests are meant for evaluating protective clothing against chemical warfare agents (NBC clothing) that manifest their effect by absorption through skin (viz., vesicant and nerve gases as discussed in Chapter 8, section 8.1). Among the various chemical agents, sulfur mustard is known to be the most penetrating in nature, and any permeable or impermeable fabric found to be effective against it ought to give a better or the same degree of protection against other members of the category. There is a lot of variation in the test methods of various countries. Prints Maurits Lab. TNO, The Netherlands, have developed and standardized various test methods, keeping in view all possible modes of exposure. These methods find widest acceptance. Two methods most commonly used for evaluation are discussed below.

9.17.1 TESTING FOR PROTECTION OF PERMEABLE CLOTHING AGAINST MUSTARD GAS VAPOR

This test is meant for vapor challenge of permeable clothing. One and one-half cm^2 of a complete clothing assembly (NBC overgarment, combat cloth, and underwear) positioned in a glass cell is exposed to an airstream of 5 m/s perpendicular to the fabric. The air (5400 L/h) is contaminated with mustard gas vapor in a concentration of 20 mg/m^3. The whole system is at room temperature (20–22°C) and at a RH of 30–80%. Through the underside, air is sucked at a speed of about 0.3–0.5 cm/s depending on the resistance to air of the complete assembly. The amount of mustard gas penetrated is collected in a bubbler from this airstream. The solvent used in the bubbler to trap the mustard gas vapor is either methylisobutylcarbinol (1–2 ml), when the bubbler is exchanged automatically every hour, or diethylsuccinate (1 ml) when only one bubbler is used in 6 hrs. The amount of mustard gas collected in the solvent is determined by gas liquid chromatography with flame photometric detector. The sample is considered to have passed the test if the collected amount of mustard in 6 hrs is <500 mg/min/m^3.

9.17.2 TESTING FOR PROTECTION OF CLOTHING AGAINST LAID DOWN MUSTARD GAS DROPS

This test is for testing the effectiveness of both permeable and imperme-able clothing against liquid challenge. A complete assembly combined with a polyethylene film (0.015 mm) is positioned on a horizontally oriented glass cell. The specimen is fixed with a glass ring of 3–5 mm height and rubber springs. The exposed surface area of the sample is 1.5 cm^2. There is a flow of air parallel to the surface of the test specimen of 0.5 m/s. A droplet of 1 μL mustard gas is placed onto the outer fabric corresponding with a contamination density of 8.3 g /m^2. A flow of air (L/h) underneath the polyethylene film transports the penetrated mustard gas vapor to a bubbler. The vapor is trapped in a solvent as in test 1. The sample is considered to have passed if the amount collected in 6 hrs is <4 μg/cm^2.

9.18 RESISTANCE TO PENETRATION BY BLOOD-BORNE PATHOGENS (ASTM F 1671-97 b)

Protective clothing is required for workers in the healthcare profession to protect them from microorganisms in the body fluids of patients. This is par-ticularly necessary for blood-borne viruses that cause hepatitis (hepatitis B and hepatitis C viruses) and acquired immune deficiency syndrome (human immun-odeficiency virus, HIV). This test method assesses the effectiveness of materials for protective clothing used for protection of the wearer against contact with blood-borne pathogens using a simulant microbe suspended in a body fluid simulant. The test equipment essentially consists of a penetration cell ($\sim$60 mL capacity), which is filled with a bacteriophage (a virus that infects bacteria), challenge suspension, and is then pressurized by air. The test specimen (mate-rial of protective clothing) acts as a barrier restraining the challenge suspension. Any penetration of the challenge suspension through the test specimen is ob-served visually at the other side of the specimen from the viewing side of the cell, and viral penetration is estimated by microbiological assay.

The challenge suspension consists of ϕX-174 bacteriophage lysate in a nu-trient broth. The ϕ-X-174 virus is not pathogenic to humans, but due to its size, similarity serves as a simulant of blood-borne pathogens including hepatitis B, hepatitis C, and HIV. The nutrient broth acts as a body fluid simulant having a surface tension of 0.042 ± 0.002 N/m. The pressure time sequence specified for the test is 0 kPa for 5 min, followed by 13.8 kPa for 1 min and 0 kPa for 54 min. Any liquid penetration on the other side of the test specimen during the test indicates failure of the sample. After the test period, the outer side of the test specimen is rinsed with a sterile nutrient broth and assayed for test virus by standard procedure. The sample is considered to have passed the test if no ϕ-X-174 virus is detected in the assay [<1 plaque forming unit (PFU)/ml].

9.19 ELECTRICAL RESISTIVITY OF FABRICS (AATCC 76-1995)

This test is important as electrical resistivity influences the accumulation of electrostatic charge of the fabric. Besides, this test is also useful for determining the conducting property of metal-coated fabric. The resistance is measured by a resistance meter. Two rectangular flat metal plates of suitable size serve as electrodes. Alternatively, two concentric ring electrodes of spacing suitable to the material can be used. The size of the test specimen should be such as to accommodate the width/diameter of the electrodes. After proper conditioning of the test specimen, the electrodes of the resistant meter are placed on it, ensuring firm contact. The resistance is measured in both length and width directions after steady state is reached, on passage of current. The lower reading in each direction is recorded. The resistivity R in ohms per square is calculated as follows. For parallel electrodes, $R = O \times W/D$ (O = measured ohms, W = width of specimen, D = distance between the electrodes). For concentric electrodes, $R = 2.73 O / \log r_0/r_1$ (r_0 and r_1 are outer and inner radius of electrodes, respectively).

9.20 REFERENCE

1. Laboratory methods for evaluating protective clothing system against chemical agents, Mary Jo Waters, Report no. CRDC-SP 84010, CRDC, Aberdeen Proving Ground, MD, U.S.A, 1984.

Properties of Some Common
Polymer Coatings[3]

Butyl rubber (isobutene-isoprene copolymers)   Good resistance to heat aging, oxidation, UV light, ozone, and general chemical attack. Low permeability to gases. Servicable temperature range −50 to +125°C. Difficult to seam. Low to moderate cost.

Hypalon (chlorosulfonated polyethylene)   Similar to neoprene. Relatively poor low temperature resistance. Moderate cost.

Natural rubber (polyisoprene)   Good tensile strength and flexibility. Tear strength and abrasion resistance improved by reinforcing fillers (e.g., carbon black). Insoluble in all organic liquids when vulcanized, but highly swollen by hydrocarbons and chlorinated solvents. Unaffected by dilute acids, alkalis, and water. Susceptible to oxidation; less so to ozone. Contains 2–4% of protein, which enhances susceptibility to biodegradation. Servicable temperature range −55 to +70°C. Sewn or glued seams required. Moderate cost.

Neoprene (polychloroprene)   Good mechanical properties. Resistant to most chemicals and organic liquids; swollen by chlorinated and aromatic solvents. Excellent weathering properties. Inferior low temperature properties to those of natural rubber. Upper temperature limit about 120°C. Low to moderate cost.

Nitrile rubber (acrylonitrile-butadiene copolymers)   Similar to natural rubber except for improved resistance to swelling in organic liquids and improved resistance to heat, light, and oxidative aging. Moderate cost.

PTFE (polytetrafluoroethylene)   Exceptional resistance to chemicals, solvents, heat, oxidation, weathering, and microorganisms. Excellent electrical and nonstick properties. Difficult to seam. Serviceable temperature range −70 to +250°C. Very high cost.

PU (polyurethanes)   Very variable compositions; properties range from hard, inflexible plastics to soft, elastic coatings. Plasticizers not required. Some grades have good resistance to fuels and oils. Excellent strength and resistance to tearing and abrasion. Thermoplastic grades available. Moderate to high cost.

[3]Adapted with permission from G. R. Lomax Textiles, vol. 14, no. 2. 1985 © Shirley Institute, U.K.

219

PVC (polyvinyl chloride)   Naturally rigid material; requires careful formulating to produce durable, flexible coatings. High plasticizer content (up to 40% by weight). Good chemical properties, although solvents tend to extract plasticizers and stiffen the polymer. Good weathering properties and flame resistance. Poor low temperature performance, unless special plasticizers are used. Thermoplastic and can therefore be seamed by hot air, radio-frequency, and ultrasonic welding techniques. Low cost.

PVDC (polyvinylidene chloride)   Similar to PVC. Better flame resistance. Low permeability to gases. Low to moderate cost.

SBR (styrene butadiene rubber)   Similar to natural rubber except for improved flex and abrasion resistance, particularly under hot, dry conditions. Inferior tear resistance and serviceable temperature range. Resistant to biodegradation. Moderate cost.

Silicone rubbers (polysiloxanes)   Inferior mechanical properties to normal rubbers. Resistant to most chemicals except concentrated acids and alkalis. Resistant to oxidation, aging, and microorgansims. Relatively high permeability to gases. Serviceable temperature range −60 to 200°C. Tasteless, odorless, and physiologically inert. Difficult to seam. High cost.

Typical Formulation of Coating Compounds

COMPOSITE COATING FOR UPHOLSTERY FABRIC[4]

BASE COAT

PVC polymer: (E, K value 68–70)	100
Stabilizer: liquid, Ca/Zn containing (e.g., Irgastab CZ 57)	1.5–3.0 phr
Costabilizer: epoxidized soya bean oil	6.0–8.0 phr
Plasticizer: DOP	85 phr
Filler: whiting	20 phr

INTERMEDIATE (EXPANDED) COAT

PVC polymer (E, K value 68–70)	100
Stabilizer/activator: liquid, Zn containing (e.g., Irgastab ABC2)	1.5–2.5 phr
Costabilizer: epoxidized soya bean oil	6.0–8.0 phr
Plasticizers: DOP	45 phr
BBP	30 phr
Blowing agent: azo dicarbonamide (paste 1:1 in DOP)	2.5–4.5 phr
Filler: whiting	5 phr

TOP COAT

PVC polymer (E, K value 70–72)	100
Stabilizer: liquid Ba/Cd/Zn Complex (e.g., Irgastab BC 206)	1.5–2.5 phr
Costabilizer: epoxidized soya bean oil	5.0 phr
Plasticizer: DOP	52 phr

[4]Reproduced with permission from *PVC Plastics* by W. V. Titow. © Kluwer Academic Publishers, the Netherlands.

Pigment: TiO$_2$ 0.0–3.0 phr
Filler: whiting   0.0–10.0 phr
Colorant: as required

POLYCHLOROPRENE COMPOUND FOR FLOATS, RAFTS, ETC.—FABRIC POLYAMIDE (Courtesy India Waterproofing and Dyeing Works, 13 Brabourne Road, Calcutta, India)

(1) Bayprene 110	100.0 (polychloroprene rubber) Bayer
(2) Magnesia	4.0
(3) Stearic acid	0.5
(4) MBTS	1.0 (dibenzthiazyl disulfide)
(5) Nonox DN	1.5 (phenyl-β-naphthyl amine)
(6) Accinox 4010 NA	0.5 (N-isopropyl N-phenyl-p-phenylene diamine)
(7) FEF black	25.0 (fine extrusion furnace black)
(8) Silica	20.0
(9) DBP	8.0 (dibutyl phthalate)
(10) Zinc oxide	5.0
(11) NA 22	0.75 (ethylene thiourea)

RUBBER COMPOUND FOR POTABLE WATER TANKS—FABRIC POLYAMIDE (Courtesy India Waterproofing and Dyeing Works, 13 Brabourne Road, Calcutta, India)

(1) Hypalon 45	100.0 (chloro sulfonated polyethylene) Dupont
(2) China clay	100.0
(3) Magnesia extra light	2.0
(4) SRF black	0.5 (semi-reinforcing furnace black)

BUTYL RUBBER COMPOUND FOR IMPERMEABLE FLAME RETARDANT PROTECTIVE CLOTHING—FABRIC POLYAMIDE (Courtesy DMSRDE, Kanpur, India)

(1) BIIR	100.0 (bromobutyl rubber 2244—Polysar)
(2) CR	20.0 (neoprene WM1—Dupont)
(3) Chlorinated paraffin wax	5.0 (58% chlorine content)
(4) Stearic acid	1.0
(5) PBN	1.0 (phenyl-β-naphthyl amine)
(6) Zinc oxide	10.0

(7) Antimony oxide	10.0
(8) Saytax	20.0 (decabromo diphenyl ether)
(9) Chlorinated polyethylene	5.0
(10) Magnesium oxide	4.0
(11) Sulfur	1.5
(12) TMTD	1.5 (tetramethyl thiuram disulfide)
(13) MBTS	1.5 (dibenzyl thiazyl disulfide)
(14) ZDC	0.5 (zinc diethyl dithiocarbamate)

POLYURETHANE FORMULATION FOR TRANSFER COATING (Courtesy M/S Entremonde Polycoater Ltd., Mumbai, India)

Solution A top coat	Impranil C granules 1.0
	Methyl ethyl ketone 2.5
Solution B tie coat	5% Imprafix TH in solution A

Impranil C, an aromatic polyester polyurethane (Bayer)
Imprafix TH, a cross-linking agent (Bayer)

ALIPHATIC POLYURETHANE FORMULATION FOR DIRECT COATING OF TOP COAT (Courtesy M/S Entremonde Polycoater Ltd., Mumbai, India)

SU 5001	1.0
Isopropyl alcohol + toluene (1:1)	0.5
Pigments if required	as desired

SU5001 is an aliphatic polyester polyurethane of Stahl GB U.K

TYPICAL THERMOPLASTIC POLYURETHANES FOR HOT-MELT COATING (Courtesy M/S Entremonde Polycoater Ltd., Mumbai, India)

a. BF Goodrich 54630 (polyether)
b. BF Goodrich 54620 (polyester)
Resins can be pigmented if necessary

Index